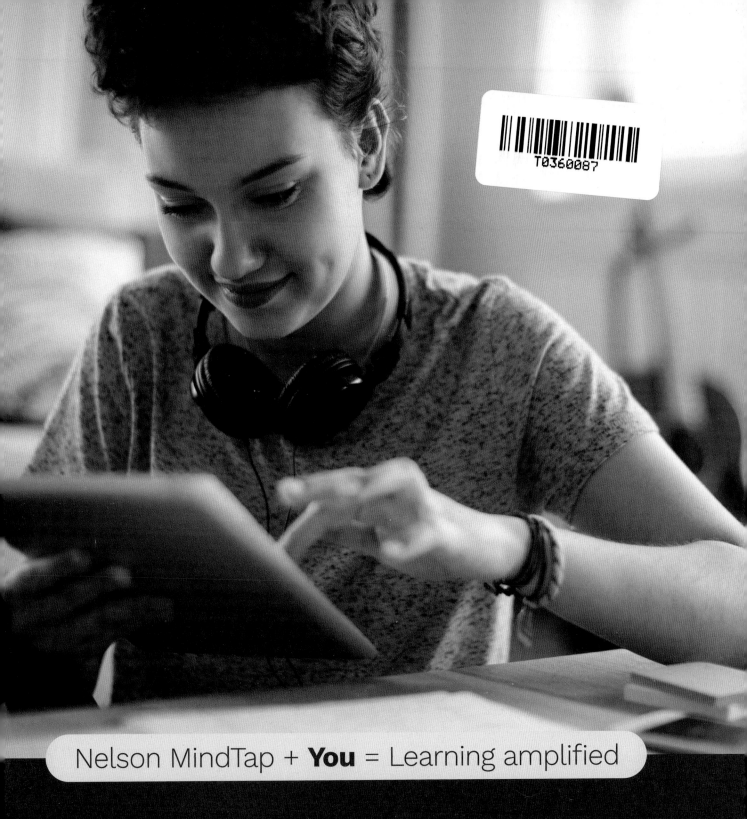

Nelson MindTap + **You** = Learning amplified

"I love that everything is interconnected, relevant and that there is a clear learning sequence. I have the tools to create a learning experience that meets the needs of all my students and can easily see how they're progressing."

— **Sarah,** Secondary School Teacher

Nelson Science Year 8 Queensland Student Book
1st Edition
Debra Smith
Christopher Huxley
Lisa McPherson
Jane Wright
Stephen Zander
ISBN 9780170463027

Series publisher: Catherine Healy
Publisher: Catriona McKenzie
Project editor: Alan Stewart
Editor: Kelly Robinson
Series text design: Leigh Ashforth
Series cover design: Leigh Ashforth
Series designer: Linda Davidson
Permissions researcher: Debbie Gallagher, Brendan Gallagher
Production controller: Bradley Smith
Typeset by: MPS Limited

Any URLs contained in this publication were checked for currency during the production process. Note, however, that the publisher cannot vouch for the ongoing currency of URLs.

Acknowledgements
Chapter 4
pages 90–91 (timeline): top left to right: Wellcome Collection. Public Domain Mark, Alamy Stock Photo/SPCOLLECTION, Alamy Stock Photo/Pictorial Press Ltd, Alamy Stock Photo/Science History Images; bottom left to right: Getty Images/De Agostini; Getty Images/Archive Photos; Alamy Stock Photo/Painters; Getty Images/Hulton Archive/Imagno
page 96 (table): column 2, top to bottom: Shutterstock.com/Wojmac, Shutterstock.com/Tatiana Ivleva, Shutterstock.com/macrowildlife; column 3, top to bottom: <TBC – new AW 05470>, <TBC – new AW 0471>, <TBC – new AW 0472>; column 4, top to bottom: Alamy Stock Photo/Science Photo Library, Shutterstock.com/Bjoern Wylezich, Getty Images Plus/E +/AndreasKermann
Chapter 5
page 127 (Question 4): top row left to right: Shutterstock.com/Natasa Re, Science Photo Library/SCIENCE SOURCE/CHARLES D. WINTERS; middle row left to right: Shutterstock.com/Christos Siatos, Science Photo Library/TURTLE ROCK SCIENTIFIC/SCIENCE SOURCE; bottom row: Shutterstock.com/PHOTO FUN

For product information and technology assistance,
in Australia call **1300 790 853**;
in New Zealand call **0800 449 725**

For permission to use material from this text or product, please email
aust.permissions@cengage.com

National Library of Australia Cataloguing-in-Publication Data
A catalogue record for this book is available from the National Library of Australia.

Cengage Learning Australia
Level 5, 80 Dorcas Street
Southbank VIC 3006 Australia

Cengage Learning New Zealand
Unit 4B Rosedale Office Park
331 Rosedale Road, Albany, North Shore 0632, NZ

For learning solutions, visit **cengage.com.au**

Printed in China by 1010 Printing International Limited.
1 2 3 4 5 6 7 26 25 24 23 22

nelson science.

8

Debra Smith
Christopher Huxley
Lisa McPherson
Jane Wright
Stephen Zander

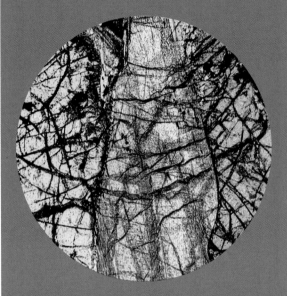

This cover image is a micrograph of peridotite, a type of igneous rock thought to make up most of Earth's mantle.

A micrograph is a photo that is taken with a microscope. This particular photo was taken at a magnification of 40 using polarised light, a technique used to enhance the contrast and improve the quality of the picture. Scientists use polarised light microscopy to identify and study solid substances such as rocks and crystals.

QLD
Australian Curriculum

FIRST NATIONS AUSTRALIANS GLOSSARY

Country/Place

Spaces mapped out that individuals or groups of First Nations Peoples of Australia occupy and regard as their own and that have varying degrees of spirituality. These spaces include lands, waters and sky.

Cultural narrative

A broad term that encompasses any cultural expression that includes (but is not limited to) knowledge and community values that are central to the identity of a particular group of First Nations Peoples.

Cultural narratives can hold information about almost anything, such as the origins of life, or can teach people about acceptable behaviour and rules, such as caring for Country.

They can take the form of songs, stories, visual arts or performances. 'Cultural narrative' is a more accurate and respectful term than 'myth', 'story' or 'fable'; terms that often diminish their importance.

First Nations Australians

'First' refers to the many nations/cultures who were in Australia before British colonisation. This a collective term that refers to all Aboriginal Peoples and Torres Strait Islander Peoples. The term 'Indigenous Australians' is also used to refer to First Nations Australians.

Nation

A self-governed community of people based on a common language, culture and territory.

Peoples and Nations

We use the plural for these terms because First Nations Australians do not belong to one nation/culture. There are many distinct Peoples and Nations. Also, some Nations consist of distinct clans or groups, so are referred to as Peoples.

Contents

First Nations Australians glossary iv
Acknowledgement of Country v
Authors and contributors viii
Nelson Science Learning Ecosystem ix
How to use this book x

1 CELLS

Chapter map 2
Big science challenge #1 3
1.1 Cell theory 4
1.2 Microscopes 6
1.3 Cells 10
1.4 Animal cells 14
1.5 Plant cells 16
1.6 Specialised cells 18
1.7 Stem cells 22
1.8 Science as a human endeavour 24
1.9 Science investigations 25
Review 28
Big science challenge project #1 29

2 ANIMAL SYSTEMS

Chapter map 30
Big science challenge #2 31
2.1 Requirements of animals 32
2.2 Body organisation 34
2.3 Digestive system 36
2.4 Respiratory system 42
2.5 Cardiovascular system 44
2.6 Urinary system 48
2.7 Disorders of body systems 50
2.8 First Nations science contexts 52
2.9 Science as a human endeavour 54
2.10 Science investigations 55
Review 58
Big science challenge project #2 59

3 PLANT SYSTEMS

Chapter map 60
Big science challenge #3 61
3.1 Requirements of plants 62
3.2 Photosynthesis 64
3.3 Systems in plants 66
3.4 Water transport 68
3.5 Sugar transport 70
3.6 Control of gases 72
3.7 Specialised structures 74
3.8 First Nations science contexts 77
3.9 Science as a human endeavour 79
3.10 Science investigations 80
Review 82
Big science challenge project #3 83

4 CLASSIFYING MATTER

Chapter map 84
Big science challenge #4 85
4.1 Atoms and elements 86
4.2 The changing concept of the atom 90
4.3 The periodic table 94
4.4 Molecules and compounds 98
4.5 Models and chemical formulas 102
4.6 Mixtures: solutions, suspensions and colloids 104
4.7 Science as a human endeavour 108
4.8 Science investigations 109
Review 113
Big science challenge project #4 115

5 CHEMICAL AND PHYSICAL CHANGE

Chapter map 116
Big science challenge #5 117
5.1 Physical change 118
5.2 Chemical change 122
5.3 Evidence of chemical change 125
5.4 Energy change in a reaction 128
5.5 Properties and uses of substances 131

5.6	First Nations science contexts	133
5.7	Science as a human endeavour	136
5.8	Science investigations	137
	Review	140
	Big science challenge project #5	143

6 THE ROCK CYCLE

	Chapter map	144
	Big science challenge #6	145
6.1	Rocks and minerals	146
6.2	Igneous rocks	150
6.3	Weathering and erosion	154
6.4	Sedimentary rocks	158
6.5	Metamorphic rocks	160
6.6	The rock cycle	164
6.7	Fossils	166
6.8	First Nations science contexts	170
6.9	Science as a human endeavour	173
6.10	Science investigations	174
	Review	177
	Big science challenge project #6	179

7 PLATE TECTONICS

	Chapter map	180
	Big science challenge #7	181
7.1	Structure of Earth	182
7.2	Plate tectonics	184
7.3	Plate boundaries	188
7.4	Continental drift	192
7.5	Seafloor spreading	194

7.6	Earthquakes and volcanoes	198
7.7	First Nations science contexts	202
7.8	Science as a human endeavour	204
7.9	Science investigations	206
	Review	209
	Big science challenge project #7	211

8 ENERGY

	Chapter map	212
	Big science challenge #8	213
8.1	Kinetic energy	214
8.2	Potential energy	218
8.3	Energy transfer	222
8.4	Energy transformation	226
8.5	First Nations science contexts	230
8.6	Science as a human endeavour	233
8.7	Science investigations	234
	Review	236
	Big science challenge project #8	237

Glossary		238
Index		248

Authors and contributors

Lead author

Debra Smith

Contributing authors

Jane Wright

Christopher Huxley

Lisa McPherson

Stephen Zander

Consultants

Joe Sambono

First Nations curriculum consultant

Dr Silvia Rudmann

Digital learning consultant

Judy Douglas

Literacy consultant

Science communication consultants

Additional science investigations

Reviewers

Anna Davis, Pete Byrne, Faye Paioff

Teacher reviewers

Aunty Gail Barrow, Nicole Brown, Christopher Evers, Associate Professor Melitta Hogarth, Carly Jia, Jesse King, Dr Jessa Rogers, Theresa Sainty

Reviewers of First Nations Science Contexts pages

9780170463027

nelson science. Learning Ecosystem

Nelson Science 8 caters to all learners

Nelson Cengage has developed a **Science Learning Progression Framework**, which is the foundation for Nelson's Science 7–10 series. An editable version is available on Nelson MindTap.

Reinforce
Nelson MindTap provides a wealth of differentiated activities and resources to meet the needs of all students

LEARN

Evaluate prior knowledge
Students complete a quiz to test their prior knowledge

LEARN

nelson science.
Nelson MindTap

Engage
Each chapter showcases fascinating, real-world science in action, while our hands-on activities, short videos and fun interactives keeps students engaged.

LEARN

Assess
Allocate and grade assessments using our differentiated end-of-topic tests and summative portfolio assessment tasks in Nelson MindTap.

LEARN

Practise
Our differentiated, scaffolded activities and investigations allow all learners to build essential skills and knowledge

LEARN

✦ Nelson MindTap

A flexible and easy-to-use online learning space that provides students with engaging, tailored learning experiences.

- Includes an eText with integrated activities and online assessments.
- Margin links in the student book signpost multimedia student resources found on Nelson MindTap.

Video activity
Cells

For students:

- Short, engaging videos with fun quizzes that bring science to life
- Interactive activities, simulations and animations that help you develop your science skills and knowledge
- Content, feedback and support that you can access as you need it and which allows you to take control of your own learning

For teachers:

- 100% modular, flexible courses let you adapt the content to your students' needs
- Differentiated activities and assessments can be assigned directly to the student, or the whole class.
- You can monitor progress using assessment tools like Gradebook and Reports.
- Integrate content and assessments directly within your school's LMS.

How to use this book

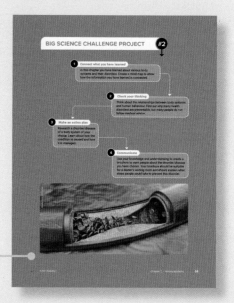

Big science, real context: The opening page begins the chapter by placing the science topic into a real-life context that is both interesting and relevant to students' lives.

Think, do, communicate: You are encouraged to reflect on and apply your learning to a set of activities, which allows you to make meaningful connections with the content and skills you have just learned.

Learning modules: Content is chunked into key concepts for effective teaching and learning.

Learning objectives: Clear, concise objectives give you oversight of what you are learning and set you up for success.

Key words: These are defined the first time they appear.

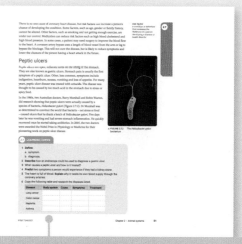

Learning check: These are engaging activities to check your understanding. Activities are presented in order of increasing complexity to help you confidently achieve the module's learning objectives. **Bolded** cognitive verbs help you clearly identify what is required of you. Activities are presented in order of increasing complexity.

Science as a Human Endeavour: Elaborations are explicitly addressed with interesting, contemporary content and activities.

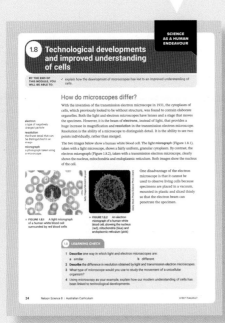

First Nations science contexts: This content was developed in consultation with a First Nations Australian curriculum specialist. It showcases the key Aboriginal and Torres Strait Islander History and Cultural Elaborations, with authentic, engaging and culturally appropriate science content.

Activities: Activities are open-ended and often hands on, helping you understand the connections between First Nations cultures and histories and science.

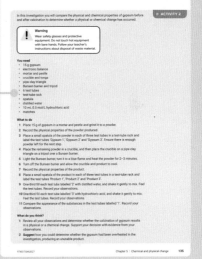

Science skills in focus: Each chapter focuses on a specific science investigation skill. This is explained and modelled with our Science skills in a minute animation, before you put it into practice in a science investigation. The science skill is reinforced with our Science skills in practice digital activities.

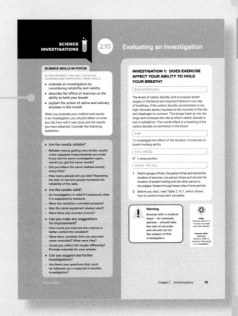

Investigations: Practise and reinforce good scientific method through fit-for-purpose, hands-on science investigations.

1

Cells

1.1 Cell theory (p. 4)

Cells are the basis of life.

1.2 Microscopes (p. 6)

The development of the microscope has enabled us to increase our knowledge of cells.

1.3 Cells (p. 10)

The basic structure of a cell changes depending on the organism it is found in and the job it does.

1.4 Animal cells (p. 14)

Different animal cells have different sizes, shapes and structures depending on their role.

1.5 Plant cells (p. 16)

Different plant cells have different sizes, shapes and structures depending on their role.

1.6 Specialised cells (p. 18)

Specialised cells have a range of different structures that relate closely to their different functions.

1.7 Stem cells (p. 22)

Stem cells are classified according to their source in the organism, and this determines their properties and value in research and medicine.

1.8 SCIENCE AS A HUMAN ENDEAVOUR: Technological developments and improved understanding of cells (p. 24)

The development of different types of microscopes has led to an improved understanding of cells.

1.9 SCIENCE INVESTIGATIONS: Science posters (p. 25)

1 Using a light microscope to see and draw cells
2 Preparing a wet mount and using a stain

9780170463027

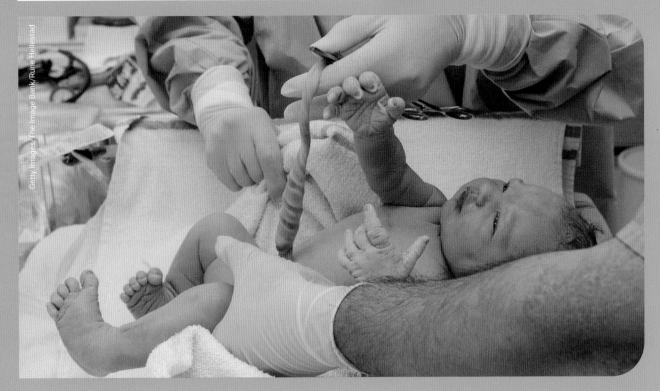

Umbilical cord blood was once discarded as waste material but is now known to be a useful source of blood stem cells. It is easy to collect, with no risk to the mother or baby. Some cord blood may be stored privately at the parents' expense, but now doctors strongly support the donation of cord blood to public blood banks.

What is cord blood used for? Why is public storage better than private storage? How can you convince parents to donate cord blood to a public blood bank?

▶ **Are you up for the challenge of learning about cord blood and stem cells?**

▲ **FIGURE 1.0.1** Umbilical cord blood is a useful source of blood stem cells.

#1 SCIENCE CHALLENGE ACCEPTED!

At the end of this chapter, you can complete Big Science Challenge Project #1. You can use the information you learn in this chapter to complete the project.

Assessments
- Prior knowledge quiz
- Chapter review questions
- End-of-chapter test
- Portfolio assessment task: Project

Videos
- Science skills in a minute: Science posters **(1.9)**
- Video activities: What is a cell? **(1.1)**; The cell membrane **(1.3)**; Stem cell research **(1.7)**

Science skills resources
- Science skills in practice: Writing science posters **(1.9)**
- Extra science investigations: Effects of surface area **(1.3)**; Examining plant cells **(1.3)**; Modelling diffusion **(1.3)**; Examining pond water **(1.4)**

Interactive resources
- Label: Microscope **(1.2)**; Animal cell **(1.4)**; Plant cell **(1.5)**
- Drag and drop: Different types of cells **(1.3)**; Plant cells **(1.5)**
- Crossword: Animal cells **(1.4)**

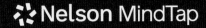 Nelson MindTap

To access these resources and many more, visit: **cengage.com.au/nelsonmindtap**

Video activity
What is a cell?

GET THINKING

What can you see in a puddle, a river or the ocean? Imagine the surprise felt by early scientists when they looked at water under their microscopes for the first time and saw tiny, living, single-celled organisms swimming around.

What is a cell?

cell
the basic structural unit of all living things

cytoplasm
a jelly-like substance that fills the inside of a cell

membrane
the thin layer that forms the outer boundary of a living cell, or of an internal cell compartment

organelle
a specialised structure in the cytoplasm of a cell that has a specific function

A **cell** is the basic building block of all living things. All organisms are made up of cells. Living things can be made up of either a single cell or many cells. All cells are composed of a jelly-like substance called **cytoplasm**, surrounded by a thin **membrane**. The membrane separates the cytoplasm from its surroundings. Specialised structures called **organelles** are found in the cytoplasm. There are about 100 trillion (100 000 000 000 000), or 10^{14}, cells in the human body.

The unfertilised ostrich egg is the biggest single cell in the world. The smallest known single cell, the bacteria *Pelagibacter ubique*, is mostly found in the oceans. The combined mass of all *Pelagibacter ubique* bacteria is more than that of all the fish in the sea. This makes it the most abundant organism on the planet.

Alan Hawk/National Museum of Health and Medicine

▲ **FIGURE 1.1.1** The microscope was invented by Hans and Zacharias Janssen. This one was made in about 1880.

Microscopes led to the discovery of cells

Because cells are so small, the history of the discovery and understanding of cells is linked to the development of microscopes. Spectacle makers Hans and Zacharias Janssen, a Dutch father and son, invented the first microscope in 1590. They put several magnifying lenses inside a tube and discovered that the lenses made objects look much larger than they were (Figure 1.1.1).

English scientist Robert Hooke made significant improvements to the basic design of this first version of the microscope. He used microscopes to examine household items such as cloth and the point of a needle. Robert Hooke was the first person to use the term 'cell' in his book *Micrographia*, published in 1665. When examining dried cork, he thought that the box-shaped structures he saw through the microscope looked like small rooms, which were often referred to as 'cells' at the time (Figure 1.1.2).

Less than 10 years later, Antonie van Leeuwenhoek used the microscopes he developed to study organisms in pond water. His microscopes were able to magnify objects very clearly and enabled him to accurately draw yeast, bark, bacteria, blood cells and sperm (Figure 1.1.3).

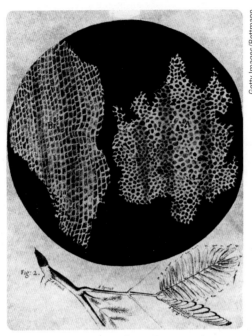

▲ **FIGURE 1.1.2** Cork cells under a simple microscope as Robert Hooke saw them in 1665

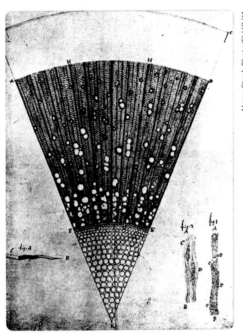

▲ **FIGURE 1.1.3** A drawing of a section of ash tree made by Antonie van Leeuwenhoek

Cell theory

Further technological developments of the microscope continued over the next 150 years. These developments helped scientists to study the internal structure of cells, their functions and how they reproduce.

In 1839, scientists Matthias Schleiden and Theodor Schwann proposed two key ideas that would become important in our knowledge of cells: that organisms are composed of cells and that the cell is the smallest unit of all organisms. Twenty years later, Rudolf Virchow showed that new cells can only arise by cell division of existing cells. These three ideas have been combined into the **cell theory**, which states the following.

- All living things are made up of cells.
- Cells are the basic building blocks of all organisms.
- Cells come from pre-existing cells; that is, new **daughter cells** are formed from the division of **parent cells**.

cell theory
the basic theory in modern cell biology that states that all living things are made up of cells, cells are the basic units of all living things, and cells form from existing cells

daughter cells
two cells that result from the division of a parent cell

parent cell
the original cell that divides to form two daughter cells

1.1 LEARNING CHECK

1 What is a cell?
2 Who was the first person to see cells?
3 **Explain** how cells were first discovered.
4 Do you think the name 'cell' is appropriate? **Explain** your reasoning.
5 **State** the cell theory.
6 **Explain** how Schwann, Schleiden and Virchow built on the work of others to formulate the cell theory.

BY THE END OF THIS MODULE, YOU WILL BE ABLE TO:

✓ identify and state the functions of the different parts of a microscope

✓ relate lens magnification to field of view

✓ relate the parts of a microscope to their function.

GET THINKING

The cells in your body are like the rooms in a house. Think about this concept as you read the information in this module, and then try to explain it in your own words.

micrometre (μm)
a unit of measurement equivalent to one-thousandth of a millimetre, or one-millionth of a metre

nanometre (nm)
a unit of measurement equivalent to one-billionth of a metre

Size of cells

Cells are usually measured in **micrometres (μm)** (one-millionth of a metre). You cannot see this size with the unaided eye, so you need a microscope to see cells (Figure 1.2.1). There are many different cells of many different sizes. Table 1.2.1 shows the sizes of some cells. Smaller specimens, such as viruses, are measured in **nanometres (nm)** (one-billionth of a metre).

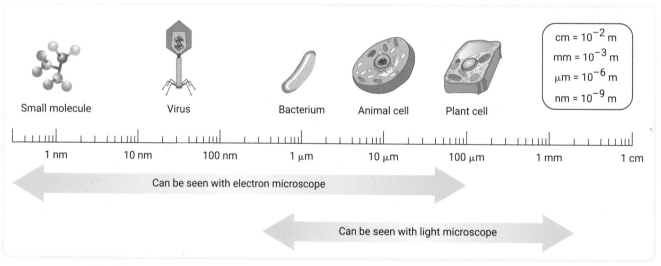

$cm = 10^{-2}\ m$
$mm = 10^{-3}\ m$
$\mu m = 10^{-6}\ m$
$nm = 10^{-9}\ m$

Small molecule   Virus   Bacterium   Animal cell   Plant cell

1 nm   10 nm   100 nm   1 μm   10 μm   100 μm   1 mm   1 cm

Can be seen with electron microscope

Can be seen with light microscope

▲ **FIGURE 1.2.1** The relative sizes of different cells

Interactive resource
Label: Microscope

▼ **TABLE 1.2.1** The different sizes of cells

Cell	Size
Hen's egg	30 mm (3 cm)

Cell	Size
Large amoeba (a single-celled organism)	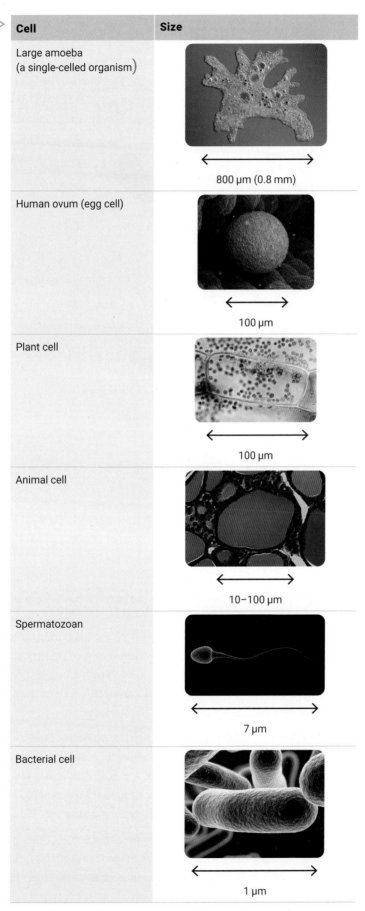

Large amoeba (a single-celled organism)
800 μm (0.8 mm)

Human ovum (egg cell)
100 μm

Plant cell
100 μm

Animal cell
10–100 μm

Spermatozoan
7 μm

Bacterial cell
1 μm

Looking at cells: the light microscope

light microscope
a microscope that uses light to view the specimen

specimen
a sample to be examined or observed

The most basic type of microscope is a **light microscope**, in which light passes through the **specimen**. It consists of a series of lenses that magnify the specimen and a light source at the base of the microscope. Look at Figure 1.2.2 and a familiarise yourself with the parts of the light microscope shown. Learn their functions from the following information.

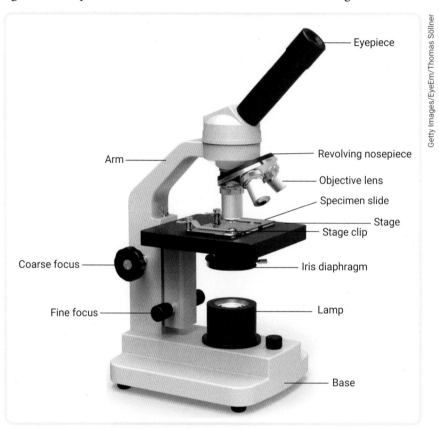

▲ **FIGURE 1.2.2**   A light microscope

stage
a flat platform that supports the slide on a light microscope

objective lens
a lens on a microscope that receives light rays from the specimen and forms an image on the eyepiece

eyepiece lens
a lens on a microscope through which the eye views the image formed by the objective lens

field of view
the diameter of the circular area that appears when you look into a microscope

magnification
the action of enlarging the apparent size of a specimen being observed

The **stage** is a flat platform. This is where the slide is placed. It has a hole in the middle to allow light, from a light source, onto the specimen. The specimen is the sample that is being examined under the microscope. Stage clips hold the specimen slide firmly in place. In microscopes with a mechanical stage, the slide is moved by turning two knobs instead of the slide being moved manually. When a slide is moved right, because of the effect of the lenses in a microscope, it will appear to move left when viewed through the microscope. Similarly, if moved down, it will seem to move up.

The two main types of lenses on light microscopes are the **objective lens** and the **eyepiece lens** (or ocular lens). There are usually three or four objective lenses on a microscope. They consist of magnifying powers of 4×, 10×, 40× and 100×, with the most powerful lens being the longest. The eyepiece, or ocular lens, is what the user looks through to view the specimen. This usually magnifies the specimen by 10×.

The **field of view** of a microscope is the diameter of the circular area that appears when you look into a microscope. As you move to a higher **magnification**, the field of view becomes smaller because you are looking at a more magnified part, meaning you are 'zooming in' further and focusing on a smaller section of the specimen.

There are two different types of adjustment knobs: the **fine focus knob** and the **coarse focus knob**. They are used to focus the microscope on the specimen by raising and lowering the stage to be closer to or further away from the objective lens. Because the fine focus knob moves the stage more slowly than the coarse focus knob, it is used at higher magnifications.

The **iris diaphragm** is located under the stage of a microscope and its main function is to regulate the amount of light that strikes the specimen.

fine focus knob
a knob that adjusts a microscope so that it focuses on the specimen by slowly raising and lowering the stage

coarse focus knob
a knob that adjusts a microscope so that it focuses on the specimen by rapidly raising and lowering the stage

iris diaphragm
a part of a microscope that regulates the amount of light that strikes the specimen

Magnification

Magnification refers to an action of enlarging the apparent size of the specimen or part of the specimen. Both the eyepiece and objective of the microscope contain magnifying lenses. When calculating the total magnification of the specimen, both magnifying lenses must be used in the calculation. For example, if the eyepiece contains a 10× lens, and the objective contains a 10× lens, then the total magnification is $10 \times 10 = 100\times$, which means the specimen will appear 100 times bigger than its actual size. Similarly, a 40× objective with a 5× eyepiece will give a total magnification of $5 \times 40 = 200\times$. When drawing a biological specimen, it is important to make a note of the total magnification at which it was viewed. This is so someone else, looking at your drawing, will know how big the specimen is in real life.

Two microscopy techniques: making a wet mount and staining a specimen

A **wet mount** is a glass slide that holds the specimen in water. This means you can look at live specimens without them drying out. Examples of specimens used in a wet mount are pond water, onion skin and thin slices of vegetable matter such as potato flesh.

A **stain**, such as iodine, can help you see the different parts of the specimen more clearly. Stains are able to enter the cell and highlight different components of the cell. This can help scientists better understand cell function.

To learn about preparing a wet mount slide and staining a specimen, refer to Module 1.9 Science investigations.

wet mount
a glass slide that holds a specimen in a liquid such as water for viewing under a microscope

stain
a dye used to colour specimens for microscopic study

1.2 LEARNING CHECK

1 What unit of measurement is used to measure a cell? How big is this unit?
2 If a slide is moved down, in what direction will it appear to move when viewed through a microscope?
3 **Calculate** the:
   a total magnification with a 10× eyepiece lens and a 40× objective lens.
   b magnification of the eyepiece lens if the objective lens is 100× and the total magnification is 400×.
4 When a person changes the objective lens from 10× to 40×, will the field of view become smaller or larger?
5 What is the purpose of using a stain on a wet mount?
6 Why is water added to a wet mount?
7 **Compare** the functions of the fine focus and coarse focus, and outline when one or the other would be used.

1.3 Cells

BY THE END OF THIS MODULE, YOU WILL BE ABLE TO:

✓ classify organisms as either unicellular or multicellular

✓ describe the difference between prokaryotic and eukaryotic cells.

GET THINKING

Skim through this module and look closely at the figures. Try to predict the difference between unicellular and multicellular organisms and prokaryotic and eukaryotic cells.

Unicellular or multicellular?

unicellular
describes a living thing consisting of a single cell

Living things that are made up of only one cell are called **unicellular**. In unicellular organisms, a single cell carries out all necessary life processes. Bacteria, amoeba, paramecium and yeast are examples of unicellular organisms (Figure 1.3.1a).

multicellular
describes a living thing consisting of more than one cell, often many cells

Many organisms are made up of large numbers of cells, and these organisms are called **multicellular**. Humans, animals and plants are examples of multicellular organisms (Figure 1.3.1b).

Although there are basic features that remain the same for all cells, the form of a cell changes depending on the organism in which it is found, and the job it does.

Video activity
The cell membrane

Interactive resource
Drag and drop:
Different types of cells

Extra science investigations
Effects of surface area
Examining plant cells
Modelling diffusion

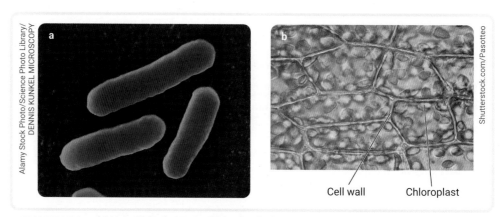

Alamy Stock Photo/Science Photo Library/ DENNIS KUNKEL MICROSCOPY

Shutterstock.com/Pasotteo

Cell wall   Chloroplast

▲ **FIGURE 1.3.1** (a) Unicellular bacteria; (b) leaf cells from a multicellular plant

The importance of being small

Most cells range from 10 to 100 µm in diameter, and in the previous module we saw that many cells are even smaller than this. There is a reason why cells are so small. A cell receives all of its requirements through its cell membrane. It also passes all of its waste out through the cell membrane. Therefore, the surface area of the cell membrane is very important – it must be large enough to allow sufficient materials through and all wastes out, for the cell to function normally.

9780170463027

The cell membrane has tiny holes, or pores, that only allow small particles to move through it. These small particles move by **diffusion**. Diffusion is when a substance, such as water, sugar or oxygen, moves from a region of high **concentration** to a region of low concentration. This means a substance in high concentration inside a cell will tend to move through the cell membrane to the outside, where the substance is found in lower concentration. The reverse is also true. Substances that are in high concentrations outside a cell will move to the inside of the cell through the cell membrane.

When substances such as water, simple sugars and oxygen are used up by the cell, their concentration decreases. This means these substances are in higher concentrations outside the cell and will move through the cell membrane to the inside of the cell. Once in the cytoplasm inside the cell, they are used in various processes that produce wastes. The wastes diffuse out of the cell from a high to a low concentration (Figure 1.3.2). This process of diffusion limits the maximum possible size of cells. As cells get larger, there is less membrane surface area compared to the volume of the cell. If the surface area of the membrane is not sufficiently larger than the volume of the cell itself, the cell will not survive. Therefore, unicellular organisms are very small, while larger multicellular organisms contain many very small cells.

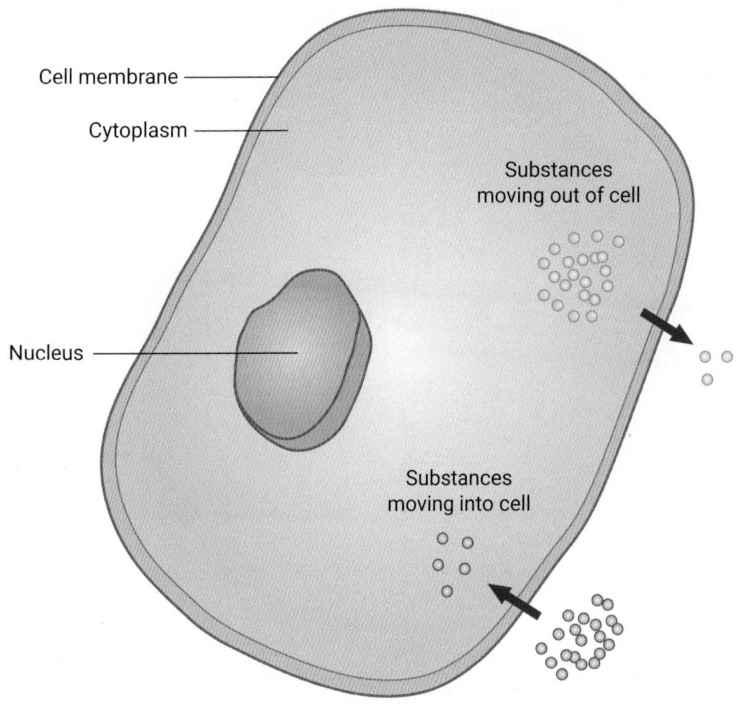

▲ **FIGURE 1.3.2**   Substances diffuse through the cell membrane from areas of high concentration to areas of low concentration.

diffusion
the movement of gas or liquid particles from a region of high concentration to a region of low concentration

concentration
the amount of solute present in a specified amount of solution

Prokaryotes

Living things are classified according to whether their cells have a **nucleus**. Organisms whose cells have a simple structure and do not have a nucleus are called **prokaryotes** (pronounced 'pro-carry-oats'). All prokaryotes are unicellular, and they have no membrane-bound organelles. These organisms are classified in the bacteria and archaebacteria kingdoms (Figure 1.3.6). Prokaryotes come in different shapes, from simple rod and spherical shapes to corkscrew spirals (Figure 1.3.3).

Although some bacteria cause disease, many bacteria do important jobs in your body and in everyday life. *Escherichia coli* (pronounced 'esh-e-reek-e-uh cole-eye', or 'E. coli' for short) lives in the intestines of humans and helps to keep your gut healthy. *Lactobacillus* and *Streptococcus* bacteria species are used to make yoghurt. Some yoghurt manufacturers add a second *Lactobacillus* strain, *Lactobacillus acidophilus*, which may have additional health benefits.

nucleus
the part of a cell that contains genetic material and is bound by the nuclear membrane

prokaryote
a unicellular organism without a nucleus

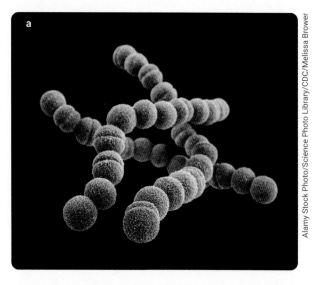

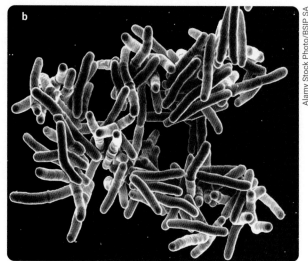

◀ FIGURE 1.3.3   Examples of prokaryotes: (a) the spherical *Streptococcus pyogenes*, which causes a 'strep' sore throat; (b) the rod-shaped (bacillus) *Mycobacterium tuberculosis*, which causes the lung disease tuberculosis; and (c) the spirochaete *Treponoma pallidum*, which causes the disease syphilis

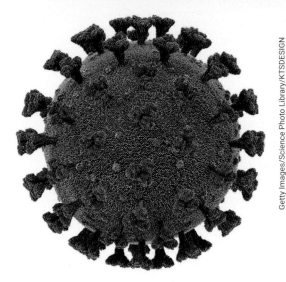

▲ FIGURE 1.3.4   A virus particle

eukaryote
an organism composed of one or more cells that contain a nucleus and membrane-bound structures

Viruses

If cells are the basic subunits of all living things, where do viruses fit in? Viruses are not classified as living things. They are much smaller than cells, ranging from 10 to 300 nanometres (nm) in diameter. Viruses do not have all the features of living organisms, such as feeding or respiration (Figure 1.3.4).

Eukaryotes

Eukaryotes (pronounced 'you-carry-oats') are organisms whose cells contain a nucleus and membrane-bound structures (Figure 1.3.5). While prokaryotic and eukaryotic cells have many features in common, the cells of eukaryotes are larger and more complex than those of prokaryotes.

The nucleus in eukaryotes is a membrane-bound organelle within the cell that keeps the nuclear material separate from the rest of the cytoplasm. Some other organelles in eukaryotes are also membrane-bound. Eukaryotes are classified in the animal, plant,

9780170463027

fungi and protist kingdoms (Figure 1.3.6). While all prokaryotes are unicellular, not all eukaryotes are multicellular. Protists and algae are examples of unicellular eukaryotes. Animals and plants are multicellular eukaryotes.

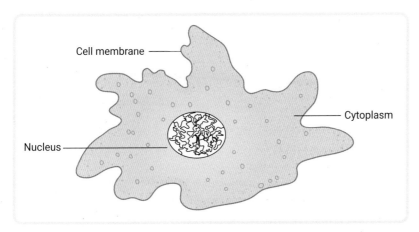

▲ FIGURE 1.3.5   A unicellular amoeba is a eukaryote because it contains a nucleus.

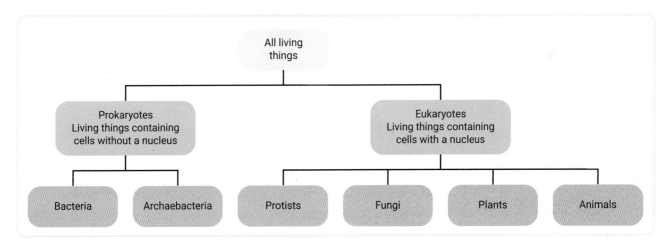

▲ FIGURE 1.3.6   Living things are classified according to whether their cells have a nucleus.

1.3 LEARNING CHECK

1 What do all prokaryotic cells have in common?
2 What two main features differentiate a eukaryotic cell from a prokaryotic cell?
3 What part of the cell forms the surface area of a cell?
4 **Explain** why oxygen diffuses into a cell and wastes diffuse out of a cell.
5 Is it true to say all prokaryotes are unicellular and all eukaryotes are multicellular? **Explain** your answer.
6 All bacteria are harmful. Is this statement true or false? **Explain** your answer.

Animal cells

GET THINKING

To maintain our energy levels, we need to eat regularly. How do our body cells obtain energy from our food, and what gas do they produce in this process?

Cellular respiration

cellular respiration
a series of chemical reactions that break down glucose into chemical energy

ribosome
the smallest organelle, which synthesises proteins

mitochondrion
an organelle in which energy is released in respiration

lysosome
an organelle that breaks down and recycles old, worn-out cell organelles

enzyme
a protein found in the body that speeds up a chemical reaction

endoplasmic reticulum
an organelle that consists of an interconnecting system of thin membrane sheets dividing the cytoplasm into compartments and channels

rough endoplasmic reticulum (RER)
endoplasmic reticulum that produces and transports proteins

smooth endoplasmic reticulum (SER)
endoplasmic reticulum that transports and produces lipids and some carbohydrates

vacuole
a membrane-bound liquid sac found inside a cell

Golgi body
an organelle that processes, packages and stores proteins and lipids

Cellular respiration is the series of chemical reactions that occur in an organism's cells to release energy from glucose, which is a simple sugar. Organisms need this energy to stay alive, grow, move, think and carry out all the vital bodily functions. In cellular respiration, glucose and oxygen react to produce energy and the waste products carbon dioxide and water. All living things undergo cellular respiration. Respiration can be summarised in this word equation:

$$\text{glucose} + \text{oxygen} \rightarrow \text{water} + \text{carbon dioxide} + \text{energy}$$

Important structures in animal cells

Organelles were introduced in Module 1.1. They are specialised cytoplasmic structures that have one or more specific functions within a cell. The smallest organelles, called **ribosomes**, are too small to be seen with a light microscope, so a more powerful microscope is used. Ribosomes carry out protein manufacture in both prokaryotic and eukaryotic cells.

Animals are eukaryotes, so their cells contain organelles with a membrane (Figure 1.4.1). These include mitochondria (singular: **mitochondrion**), which carry out respiration, and **lysosomes**, which contain **enzymes** to break down and recycle old, worn-out cell parts.

Proteins and other molecules within eukaryotic cells are transported by the **endoplasmic reticulum**, which is an interconnecting system of thin membrane sheets dividing the cytoplasm into compartments and channels. The endoplasmic reticulum can be either rough or smooth. The **rough endoplasmic reticulum (RER)** is covered in ribosomes that produce proteins. Proteins made by ribosomes on the RER move directly into channels that help transport them through the cell or export them via the cell membrane. The **smooth endoplasmic reticulum (SER)** has no ribosomes. It is responsible for the transport and synthesis of lipids and some carbohydrates.

Vacuoles are liquid-filled compartments surrounded by a membrane. In animals, they are very small and numerous. In some unicellular eukaryotes, the vacuoles are responsible for the storage and removal of water from the cell. Some vacuoles store food and others store waste products.

The **Golgi body** is an organelle that processes, packages and stores proteins and lipid molecules, especially proteins to be exported from the cell. It appears as a series of flattened stacked membranes near the nucleus.

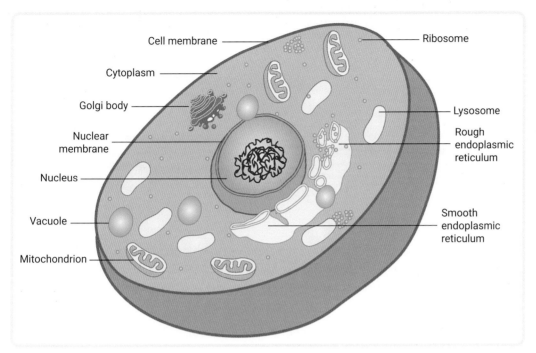

Cell membrane ——————

Cytoplasm ——————

Golgi body ——————

Nuclear membrane ——————

Nucleus ——————

Vacuole ——————

Mitochondrion ——————

—————— Ribosome

—————— Lysosome

—————— Rough endoplasmic reticulum

—————— Smooth endoplasmic reticulum

▲ **FIGURE 1.4.1** A typical animal cell

Making a model: The animal cell open sandwich

You need
- slice of bread
- knife
- jam (cytoplasm)
- chocolate biscuit (nucleus)
- poppy seeds (ribosomes)
- chocolate bits (small vacuoles)

- pieces of dried apricots skewered together with a toothpick (Golgi body)
- yellow lolly snakes sliced into thin strips (cell membrane)
- hundreds and thousands sprinkles (other organelles)

Warning

Food in the lab can be contaminated. Do not eat any of the materials in this model.

What to do
Use the ingredients listed to construct an animal cell open sandwich. Look at the animal cell diagram (Figure 1.4.1) if you need help.

What do you think?

1 **Evaluate** your model by describing its strengths and weaknesses.

2 **List** the extra ingredients you would use to create a plant cell sandwich. Share your list with the class.

1.4 LEARNING CHECK

1 **Describe** the process of cellular respiration.

2 What is an organelle?

3 **State** the functions of mitochondria and ribosomes.

4 Why is a cell membrane important to a cell?

5 **Compare** the structure and function of smooth and rough endoplasmic reticulum.

6 **Explain** the importance of cellular respiration to an animal.

Interactive resources
Crossword: Animal cells
Label: Animal cell

Extra science investigation
Examining pond water

✓ define photosynthesis

✓ describe the structure and function of key structures in plant cells.

GET THINKING

Imagine that you are eating a carrot. In what way do you think the plant cells in the carrot are different from the animal cells you learned about in the previous module?

photosynthesis
the process by which a green plant uses sunlight, water and carbon dioxide to produce glucose, which it uses for nutrition, and oxygen, which is released into the air

chloroplast
a green organelle in plant cells that contains chlorophyll and carries out photosynthesis

chlorophyll
the green pigment in chloroplasts that absorbs light for photosynthesis

cell wall
the rigid outer covering of bacterial and plant cells that surrounds the cell membrane

cellulose
a complex carbohydrate that makes up the cell walls of plants

Photosynthesis

Photosynthesis is a series of chemical reactions in which water (taken up by the roots of plants) and carbon dioxide (from the air) are converted into oxygen and glucose. Light energy from the Sun is used for this process, and some of it ends up stored as chemical energy (Figure 1.5.1).

Photosynthesis can be summarised by this word equation:

$$\text{carbon dioxide} + \text{water} \xrightarrow[\text{chlorophyll}]{\text{sunlight}} \text{glucose} + \text{oxygen}$$

The **chloroplasts** in plant cells are responsible for photosynthesis. They are green because they contain the pigment **chlorophyll** that traps the light energy, helping to convert it into chemical energy that is stored in the glucose produced. Plants also undergo cellular respiration, using their stored energy for living and for growth.

Photosynthesis is discussed further in Chapter 3.

Plant cells and their structures

All plants have a similar basic cell structure. Plant cells contain the same structures as animal cells, including ribosomes, smooth and rough endoplasmic reticulum, Golgi bodies, mitochondria, lysosomes and a nucleus. As shown in Table 1.5.1, these structures perform the same functions in plants as they do in animals. Plant cells also have three additional structures not found in animal cells: chloroplasts, a large central vacuole (as opposed to numerous small vacuoles) and a **cell wall**.

Chloroplasts are clearly visible under the light microscope and occur in almost all plant cells exposed to light.

In plants, the large vacuole is a water-filled compartment surrounded by a membrane. It stores water and chemicals and provides support, maintaining the structure of the plant by holding water that pushes out against the cell wall.

The cell wall is the rigid outer covering surrounding the cell membrane. It is made of **cellulose**. Cell walls give the plant cell shape and protection as well as strength and support. This function is important because, unlike animals, plants do not have a rigid skeleton or muscles that strengthen and support the body.

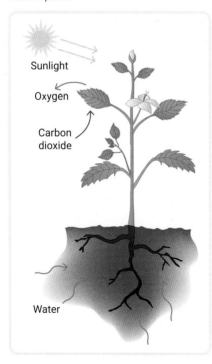

Sunlight

Oxygen

Carbon dioxide

Water

▲ **FIGURE 1.5.1**
Requirements for photosynthesis

▼ TABLE 1.5.1   Summary of the functions of organelles in plants. Those marked * are not found in animals

Structure	Function
Mitochondrion	Site of respiration
Golgi body	Processes, packages and stores proteins and lipids
Ribosome	Produces proteins
Rough endoplasmic reticulum	Produces and transports proteins
Smooth endoplasmic reticulum	Transports and produces lipids and some carbohydrates throughout the cell
Lysosome	Breaks down and recycles old, worn-out cell parts
Large central vacuole *	Stores water and chemicals
Cell wall *	Rigid outer covering of a plant cell found outside the cell membrane
Chloroplast *	Makes glucose and oxygen through the process of photosynthesis

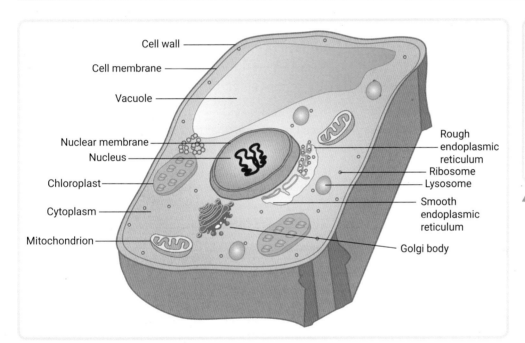

▲ FIGURE 1.5.2   A typical plant cell

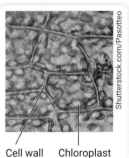

▲ FIGURE 1.5.3   Cell walls and chloroplasts are clearly visible in these plant cells.

Cell wall   Chloroplast

Shutterstock.com/Pasotteo

1.5   LEARNING CHECK

1   **Explain** the importance of photosynthesis in plants.

2   **Describe** the function of ribosomes in plants.

3   **Compare** and contrast the organelles found in animal cells and plant cells.

4   **Predict** whether it is possible to find a plant cell that does not contain chloroplasts. Explain your answer.

5   Which type of cell – plant, animal or prokaryote – do you think is the most interesting? **Explain** your answer.

Interactive resources
Drag and drop: Plant cells
Label: Plant cell

1.6 Specialised cells

BY THE END OF THIS MODULE, YOU WILL BE ABLE TO:

✓ explain why cells become specialised
✓ describe the relationship between the structure and function of a range of specialised cells.

Quiz
Specialised cells

GET THINKING

Skim read this module, paying particular attention to the figures. Think about the range of specialised shapes and structures you can see in these cells. Note down a description of the most interesting ones.

Specialisations of unicellular organisms

Unicellular eukaryotic organisms have specialisations that allow them to live successfully in their environments. For example, to cope with living in fresh water, where water constantly diffuses through the membrane into the cytoplasm, an amoeba contains contractile vacuoles (Figure 1.6.1). These vacuoles gradually fill with water and, when stretched to a certain point, they contract, expelling the water from the cell.

The amoeba also has a very flexible cell membrane that makes finger-like temporary extensions of its cytoplasm. This allows it to move by flowing across the surface of rocks and mud, and lets it surround its prey. Figure 1.6.2 shows an amoeba sending out projections of its cell membrane to feed on a smaller organism. In this way, a food vacuole is formed in the cytoplasm. Enzymes within the food vacuole gradually digest the prey and any wastes are excreted out of the cell from the same vacuole.

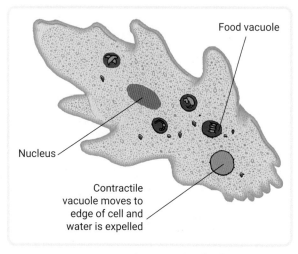

▲ **FIGURE 1.6.1** An amoeba containing food vacuoles and contractile vacuoles

▲ **FIGURE 1.6.2** An amoeba surrounding its prey before engulfing it

Specialisation in multicellular animals

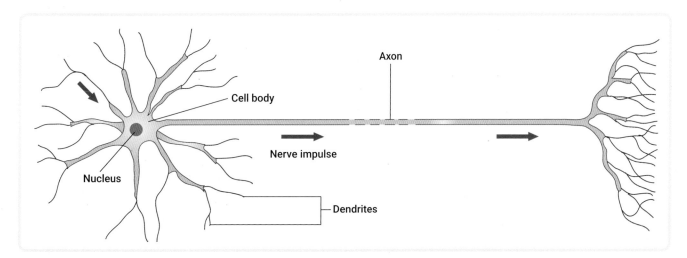

In multicellular animals, a process called **differentiation** leads to specialised cells having cellular structures and contents that allow them to carry out a specific function. Although the size, shape and chemical composition of a cell often change during differentiation, the genetic material remains unchanged.

differentiation
a biological process whereby cells of an organism become specialised

Specialised animal cells

We saw in Module 1.4 that animal cells consist of a cell membrane surrounding cytoplasm in which a nucleus, mitochondria and other organelles occur. In most animal cells, this basic plan is modified to enable the cell to carry out a specific function. For example, fat cells are specialised to store lipids for energy. The muscle cells in your heart are long and thin and contain fibres that can contract and cause your heart to beat. Muscle cells contain very large numbers of mitochondria. They carry out respiration to provide energy for muscle contraction.

Neurons, or nerve cells, are the part of the nervous system that carries messages to and from different parts of the body (Figure 1.6.3). Each neuron has a very long **axon**, or projection, from the cell body along which nerve impulses are transmitted. The axons that carry messages between your spine and feet are more than one metre long. Their branched connections at each end allow them to connect with other neurons.

neuron
a nerve cell

axon
the part of a neuron that carries the nerve impulse

Axon

Cell body

Nerve impulse

Nucleus

Dendrites

▲ **FIGURE 1.6.3**   A neuron with a cell body and an axon

Red blood cells are highly specialised cells that carry oxygen around the body in the blood. They are packed with haemoglobin, which carries oxygen molecules. During differentiation, all their organelles, including the nucleus, disappear to ensure maximum space for the haemoglobin. Their shape, like flattened discs with dips on both sides, is called **biconcave** (Figure 1.6.4). This shape provides a large surface area for the absorption of oxygen in the lungs. Red blood cells live for only 120 days. Old and damaged red blood cells end up in the liver, where they are broken down by white blood cells to enable recycling of their components.

biconcave
shaped like a flattened disc with dips on both sides

There are many different types of white blood cells in the blood. All help defend the body against disease. Some, like the macrophage in Figure 1.6.5, are responsible for the detection and destruction of invading micro-organisms. Their cell membrane contains a system of recognition receptors to detect bacteria and dead body cells. The membrane forms long, thin extensions from the main cell body. These can attach to foreign cells, which can be surrounded and destroyed.

You will learn more about red and white blood cells in Chapter 2.

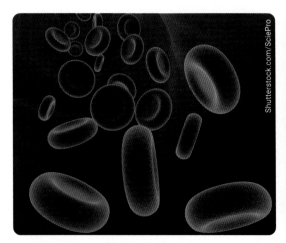

▲ **FIGURE 1.6.4** Red blood cells have a biconcave shape

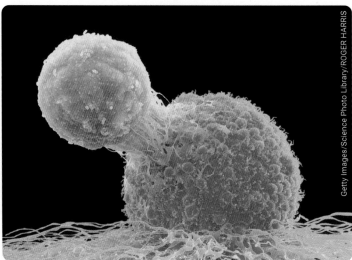

▲ **FIGURE 1.6.5** A white blood cell using long extensions to attach to a cancer cell (coloured blue)

Specialised plant cells

root hair
a long extension that provides a large surface area for the root of a plant

Plants, too, have specialised cells that perform special functions. Roots absorb water from the soil for use by other parts of the plant. Figure 1.6.6 shows a cytoplasmic projection, called a **root hair**, from the surface of the root into the soil. Thousands of root hairs, just behind the root tip, provide an enormous surface area for water absorption.

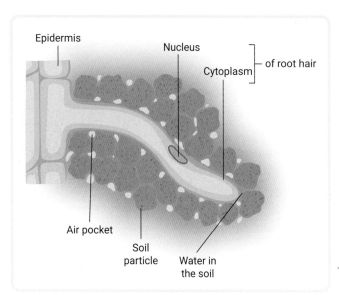

Epidermis
Nucleus
Cytoplasm ⎤ of root hair
Air pocket
Soil particle
Water in the soil

◄ **FIGURE 1.6.6**
A root hair cell

9780170463027

Guard cells are specialised crescent-shaped cells on the surface of leaves. They are arranged in pairs and form a small pore on the leaf surface. These pores, called **stomata (singular: stoma)**, allow gases such as carbon dioxide and oxygen to enter and leave the leaf by diffusion. These gases are essential for photosynthesis and respiration by plant cells. Guard cells can regulate the amount of water loss from a leaf, such as during hot days or at night, by changing their shape and causing the stomata to close (Figure 1.6.7). Guard cells contain significant numbers of chloroplasts. Sugars made from photosynthesis control the change in the shape of guard cells, to open the stomata during the day.

guard cells
cells that surround the stomata of a plant, allowing them to open and close

stomata (singular: stoma)
pores on the surfaces of leaves that allow gas exchange

1.6

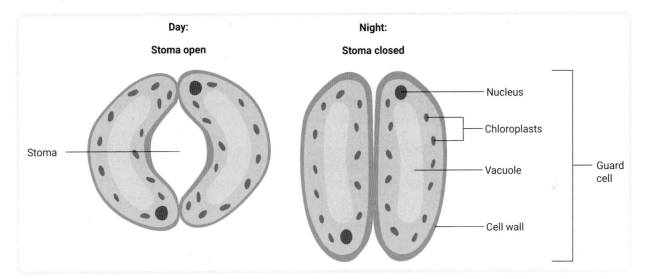

▲ **FIGURE 1.6.7**   A stoma, formed by a pair of guard cells that cause the stoma to open during the day and close at night.

You will learn more about plant root hairs, stomata and guard cells in Chapter 3.

1.6 LEARNING CHECK

1 **Define** 'specialised cell'.
2 **Describe** two specialisations of an amoeba.
3 **State** the function of a guard cell and describe how it is specialised to carry out this function.
4 **Describe** two specialisations of nerve cells and relate these to their function.
5 **Explain** why some cells are specialised.
6 Red blood cells only live for 120 days. What are the implications of this for:
    a a blood donor?
    b a blood donor recipient?
    c the blood bank?

BY THE END OF THIS MODULE, YOU WILL BE ABLE TO:

✓ relate the source of stem cells to their properties and role within an organism

✓ explain the value of stem cells in medical treatments and research.

Video activity
Stem cell research

GET THINKING

Think back to a time when you grazed your knee or cut your finger. Did your body repair itself? Where did the new cells that healed your skin come from? Jot down your ideas. Now, read the module to find out whether you were correct.

What are stem cells?

In the last module, you learned that red blood cells do not contain a nucleus and live for only 120 days. How do they get replaced? The answer is that new red blood cells are produced from **stem cells** in the bone marrow. These stem cells in bone marrow form all blood cells, including the white blood cells and platelets.

Stem cells are **undifferentiated** cells found in small numbers in all parts of an organism. They can divide indefinitely. Once these stem cells have divided, one daughter cell remains a stem cell and the other undergoes differentiation to become a specialised cell. (Figure 1.7.1)

Stem cells play an important role in replacing cells damaged by normal wear and tear, infection and disease. These cells are referred to as **adult stem cells**, although they occur naturally in people of all ages. In general, they produce the type of tissues in which they occur. For example, skin stem cells replace damaged cells in the skin. Because of this limited range, they are referred to as **multipotent** stem cells.

Stem cells can be classified according to where they come from in the body and on their ability to differentiate. In mammals, the fertilised egg and the products of the initial few cell divisions in the embryo are the only **totipotent** stem cells. This means they can develop into a complete organism or differentiate into any of its cells or tissues, including placental tissue. As the embryo develops over the next few days, the **embryonic stem cells** change from being classed as totipotent to **pluripotent**. These cells retain the ability to differentiate into any type of body cell.

stem cell
an unspecialised cell able to keep dividing, with the potential to become specialised

undifferentiated
describes a cell that has not yet developed into a specialised cell type

adult stem cell
an undifferentiated cell found everywhere in the body, which can divide to replace dead or damaged cells

multipotent
describes a cell that can differentiate into only a few cell types

totipotent
describes a cell that can differentiate into any type of cell or a complete mammal

embryonic stem cell
a cell from an embryo that is three to five days old; can divide into more stem cells or can become any type of cell in the body; it is pluripotent

pluripotent
describes a cell that has the ability to differentiate into any type of cell in the body

▲ FIGURE 1.7.1   A stem cell gives rise to more stem cells or specialised cells.

Use of stem cells in medicine and research

Once discovered, embryonic stem cells were harvested from spare embryos from fertility treatment and used in scientific research. This raised important ethical issues for some people because they believed that the use of an embryo in this way represented destruction of human life.

In 2006, a significant breakthrough occurred when scientists found a relatively simple way to reprogram any ordinary body cell to become a pluripotent stem cell. These cells are called **induced pluripotent stem cells**. As they behave in a similar way to embryonic stem cells, they can be used in the laboratory to make almost any type of body cell. This milestone overcame previous ethical problems in stem cell research (Figure 1.7.2).

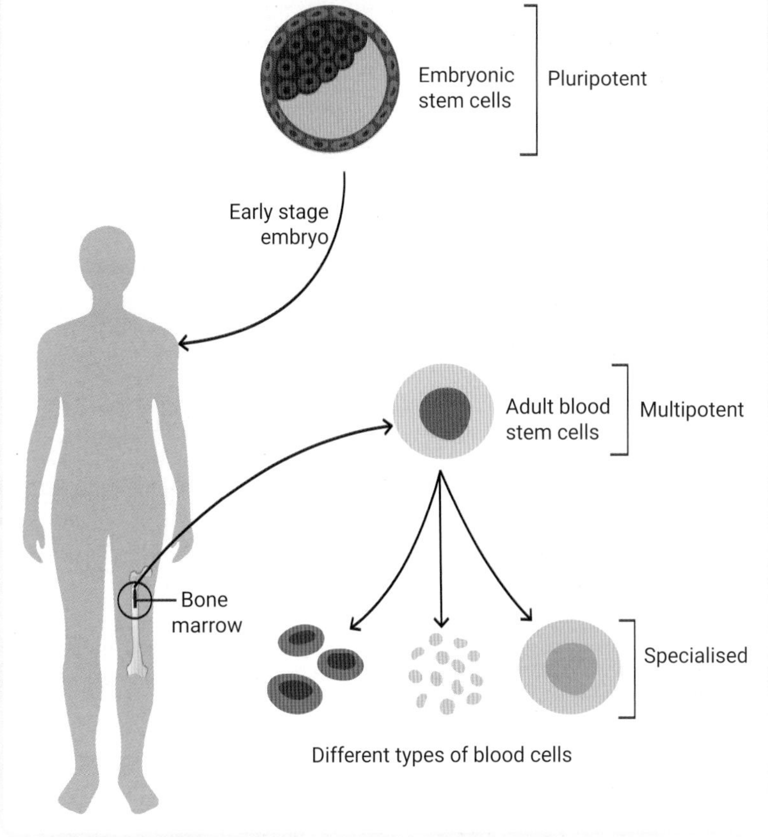

Although there is a lot of interest in the application of stem cells in medicine, the number of approved stem cell therapies is small. For the past 40 years, stem cells in bone marrow have been used to replace cells destroyed during treatment of a variety of cancers and blood diseases. Another approved application of stem cells is in creating tissue grafts to replace diseased or damaged skin and corneas (in the eye).

Scientists hope that stem cells will be able to regenerate damaged tissues, such as after an injury to the spinal cord or the death of heart muscle from a heart attack. The induced pluripotent stem cells being tested are derived from a patient's own adult cells. Being perfectly matched to the patient means there is no chance of immune rejection of the stem cells.

Another line of research is to use stem cells to replace damaged retinal cells in patients with macular degeneration. Macular degeneration is the loss of retinal cells in the eye, leading to loss of central vision.

▲ **FIGURE 1.7.2** Pluripotent stem cells can differentiate into any type of body cell; multipotent stem cells can only differentiate into the types of tissues in which they occur.

induced pluripotent stem cell
a cell that is reprogrammed from adult tissues to become pluripotent

1.7 LEARNING CHECK

1 **Name** the types of cells produced by bone marrow.
2 **Compare** and contrast the meaning of pluripotent and multipotent.
3 **Explain** how your skin repairs itself if you cut your finger.
4 What type of stem cells would be used to find a cure for a degenerative condition such as arthritis?
5 What ethical issue was involved in the use of embryonic stem cells for research?
6 Would you consider stem cell treatment if you were suffering from macular degeneration? Give your reasons.

1.8 Technological developments and improved understanding of cells

SCIENCE AS A HUMAN ENDEAVOUR

BY THE END OF THIS MODULE, YOU WILL BE ABLE TO:

✓ explain how the development of microscopes has led to an improved understanding of cells.

How do microscopes differ?

electron
a type of negatively charged particle

resolution
the finest detail that can be distinguished in an image

micrograph
a photograph taken using a microscope

With the invention of the transmission electron microscope in 1931, the cytoplasm of cells, which previously looked to be without structure, was found to contain elaborate organelles. Both the light and electron microscopes have lenses and a stage that moves the specimen. However, it is the beam of **electrons**, instead of light, that provides a huge increase in magnification and **resolution** in the transmission electron microscope. Resolution is the ability of a microscope to distinguish detail. It is the ability to see two points individually, rather than merged.

The two images below show a human white blood cell. The light **micrograph** (Figure 1.8.1), taken with a light microscope, shows a fairly uniform, granular cytoplasm. By contrast, the electron micrograph (Figure 1.8.2), taken with a transmission electron microscope, clearly shows the nucleus, mitochondria and endoplasmic reticulum. Both images show the nucleus of the cell.

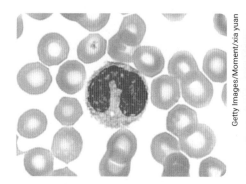

Getty Images/Moment/xia yuan

▲ **FIGURE 1.8.1** A light micrograph of a human white blood cell surrounded by red blood cells

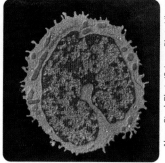

Alamy Stock Photo/Science Photo Library

▲ **FIGURE 1.8.2** An electron micrograph of a human white blood cell, showing the nucleus (red), mitochondria (blue) and endoplasmic reticulum (pink)

One disadvantage of the electron microscope is that it cannot be used to observe living cells because specimens are placed in a vacuum, mounted in plastic and sliced thinly so that the electron beam can penetrate the specimen.

1.8 LEARNING CHECK

1 **Describe** one way in which light and electron microscopes are:
   a similar.
   b different.
2 **Describe** the difference in resolution obtained by light and transmission electron microscopes.
3 What type of microscope would you use to study the movement of a unicellular organism?
4 Using microscopy as your example, explain how our modern understanding of cells has been linked to technological developments.

9780170463027

IN THIS MODULE, YOU WILL FOCUS ON LEARNING AND IMPROVING THESE SKILLS:

▶ explain how to set up and use a microscope

▶ explain how to prepare a wet mount slide

▶ use a microscope to examine and draw cells, showing the relative size of cell organelles

▶ produce a simple science poster to present the results from this investigation of cells.

What is a science poster? A science poster is like a mini, visual version of a science report. Posters usually contain images such as graphs, photos or diagrams, and not too many words.

What are they used for? Scientists usually make posters as a concise way to share their science research or data with other scientists.

What information does a science poster have? They usually have the same sections as a science report.

Video
Science skills in a minute: Science posters

Science skills resource
Science skills in practice: Writing science posters

▶ A science poster will usually include the following sections.

- Title: This is often the question you asked in your project or research.
- Introduction: State the problem you were trying to solve and the aims of the research.
- Method: Briefly explain the materials you used and the steps you took in your investigation.
- Results: Present the information you collected or the observations you made. It is best to use photographs, graphs or drawings to help people understand your results.
- Discussion and conclusion: Explain what your findings mean.
- References: Keep the list as short as possible.

INVESTIGATION 1: USING A LIGHT MICROSCOPE TO SEE AND DRAW CELLS

AIM

To view a prepared slide of human blood using a light microscope and draw a diagram of the cells observed

Warning
Carry the microscope with two hands and always use it on a level surface. Take care when using glass slides. Report any breakages to your teacher immediately.

YOU NEED

☑ prepared slide of a sample of human blood
☑ light microscope

WHAT TO DO

1 Place the slide onto the microscope stage and secure it with stage clips. Using the 10× objective lens and the coarse focus knob, bring the image into focus. Start with the stage as close to the objective lens as possible, then slowly turn the coarse focus knob to move the stage away until the slide comes into focus.

2 Next, look at the slide using the 40× and 100× objective lenses. Start from the focused position using the previous lens. Remember to only use the fine focus knob to focus.

3 Identify the red and white blood cells. Your teacher will help you do this.

4 Make a biological drawing of the sample at 100×, showing the relative sizes of the red and white cells.

WHAT DO YOU THINK?

1 What problems did you encounter while using the light microscope?

2 What are the obvious differences between red and white blood cells?

INVESTIGATION 2: PREPARING A WET MOUNT AND USING A STAIN

To prepare and look at your own slide of biological material and investigate the effect of using a stain to help you see the different parts of the cells

Warning

Take care when using a scalpel. Always cut away from yourself. Carry the scalpel in a tray with the point facing away from you. Never run when carrying a sharp object.

Always use a cutting board. Place the scalpel in the middle of the board so that it doesn't fall off the edge. Report any cuts to your teacher immediately.

Wash your hands after completing the activity. Clean up any iodine spills immediately.

YOU NEED

- ☑ light microscope
- ☑ microscope slides
- ☑ coverslips
- ☑ eyedropper
- ☑ onion
- ☑ forceps
- ☑ scalpel
- ☑ cutting board
- ☑ iodine stain in a dropper bottle
- ☑ water
- ☑ paper towel

WHAT TO DO

PART A: PREPARING THE SLIDE

1 Use the eyedropper to place one drop of water onto the middle of the microscope slide (Figure 1.9.1).

▲ FIGURE 1.9.1   Place a drop of water onto the slide.

2 Use the scalpel to cut a small piece of onion (Figure 1.9.2).

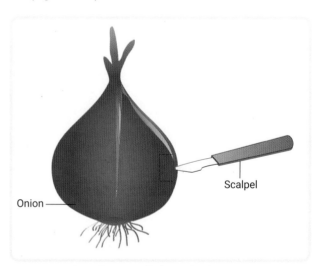

Onion

Scalpel

▲ FIGURE 1.9.2   Cut a small piece of onion.

3 Use the forceps to peel a small piece of thin membrane from the onion (Figure 1.9.3).

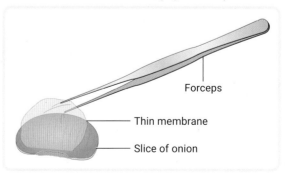

Forceps

Thin membrane

Slice of onion

▲ FIGURE 1.9.3   Peel a piece of membrane from the onion.

4 Lay the thin membrane flat on top of the drop of water on your slide (Figure 1.9.4).

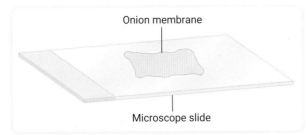

Onion membrane

Microscope slide

▲ **FIGURE 1.9.4** Lay the membrane onto the drop of water.

5 Carefully lower the coverslip. To do this without trapping air bubbles underneath it, you need to place one end of the coverslip so it rests on the slide at the edge of the water drop. Use the forceps to lower the other end of the coverslip onto the water drop and specimen as shown in Figure 1.9.5.

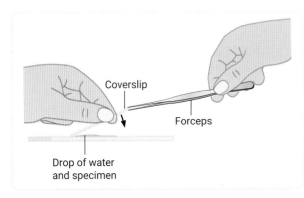

Coverslip

Forceps

Drop of water and specimen

▲ **FIGURE 1.9.5** Lower the coverslip onto the specimen.

PART B: Viewing the slide

6 Place the slide onto the microscope stage and secure it with stage clips. Using the low power (10×) objective lens and the coarse focus knob, bring the image into focus.

7 Look at the slide using 40× and 100× objectives. Remember that, when you are using the 40× and 100× objectives, you will need to use the fine focus knob to focus.

8 Make a biological drawing of the sample at an appropriate magnification.

9 Use a stain to assist you in seeing the different parts of the onion cells more clearly. Place one drop of iodine stain at the edge of the coverslip. Place a piece of paper towel on the other side of the coverslip to draw the iodine stain under the coverslip, as shown in Figure 1.9.6.

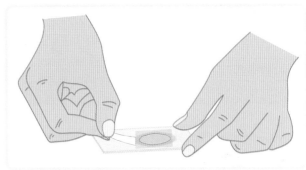

▲ **FIGURE 1.9.6** Draw the stain under the coverslip with a piece of paper towel.

10 Repeat Steps 6–8 using the slide of stained cells.

WHAT DO YOU THINK?

1 When making your own slide, what was challenging and what did you find easy?

2 What did you see under the highest magnification?

3 Record the differences and similarities you observed between an onion cell and a red blood cell (from Investigation 1).

4 What structures could you distinguish within the cells?

5 In what ways did the stain improve your ability to see parts of the cells?

CONCLUSION

1 What can you conclude about the value of microscopes in adding to your understanding of the structure of cells?

2 Produce a science poster to report on your investigation.

1 REVIEW

REMEMBERING

1 **Define** 'photosynthesis'.

2 **State** the function of the fine focus knob of a microscope.

3 What is the role of the Golgi body in a cell?

4 **State** the function of lysosomes.

5 **Describe** three features common to eukaryotic cells.

UNDERSTANDING

6 **Explain** why cells were not known about until the 1660s.

7 **Name** the organelle that controls the functioning of eukaryotic cells.

8 **Explain** why stains are often used when preparing a wet mount slide.

9 **Explain** why cells are considered to be the basis of life.

10 **Explain** why a cell membrane is important to a cell.

11 **Explain** why you would expect a muscle cell to contain more mitochondria than a skin cell.

12 If plant cells can make their own food, **explain** why they need mitochondria.

APPLYING

13 If you were given an unknown cell, **explain** how you would be able to tell whether it was:

a prokaryotic or eukaryotic.

b from a plant or an animal.

14 **Explain** why the use of stem cells has ethical considerations.

15 **Describe** the steps that you would carry out to make a wet mount slide of plant leaf cells.

ANALYSING

16 **Justify** why large organisms are multicellular.

17 **Compare** and contrast a plant leaf cell with a human muscle cell.

18 The structure of a guard cell is ideally suited to its function. Provide at least two reasons to support this statement.

EVALUATING

19 a Evaluate the role of bacteria in our world.

b Conduct a debate on the topic: Bacteria are an essential and necessary group of organisms.

20 A large cell of 0.3 mm was observed. Ribosomes were present in the cytoplasm, but no other organelles were visible. Could this be a prokaryotic or eukaryotic cell? **Justify** your answer. What other evidence would be useful in deciding?

CREATING

21 Use an online crossword generator to create a crossword for the eight parts of a light microscope.

22 Imagine you are interviewing Antonie van Leeuwenhoek or Robert Hooke. What are three questions you would ask?

23 You are given a spherical cell. **Describe** how you could redesign it so that it has a large surface area. Hint: think about what shape your cell would have to be.

24 Create an argument to persuade your classmates that it is an advantage for eukaryotic cells to have different types of organelles.

9780170463027

BIG SCIENCE CHALLENGE PROJECT #1

1 Connect what you have learned

In this chapter you have learned that stem cells play an important role in replacing cells damaged by normal wear and tear, infection and disease. Create a mind map to show how the information that you have learned about stem cells is connected.

2 Check your thinking

Find out how blood stem cells help people being treated for blood cancers such as leukaemia.

Research the importance of tissue matching in transplanting blood stem cells.

Explain the value to the community of storing umbilical cord blood in public blood banks.

3 Make an action plan

Consider the following questions during your research.

What is the problem?

What is the solution?

Who should be targeted?

What will the benefit be?

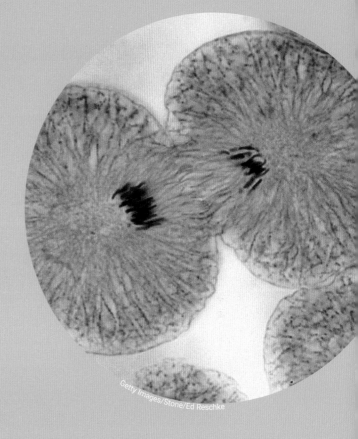

Getty Images/Stone/Ed Reschke

4 Communicate

Use your knowledge and understanding to create a promotional campaign to encourage mothers to donate their newborn baby's umbilical cord blood to a public cord blood bank.

Animal systems

2.1 **Requirements of animals** (p. 32)

All animals require water, oxygen and food for energy, growth and repair.

2.2 **Body organisation** (p. 34)

Cells, tissues and organs form systems that work together to make the functioning organism.

2.3 **Digestive system** (p. 36)

The digestive system breaks down food for absorption into the body.

2.4 **Respiratory system** (p. 42)

The respiratory system facilitates the intake of oxygen and its movement into the blood and the removal of carbon dioxide from the body.

2.5 **Cardiovascular system** (p. 44)

The cardiovascular system transports nutrients and waste to and from every cell in the body through the blood.

2.6 **Urinary system** (p. 48)

The urinary system removes toxic waste from the body.

2.7 **Disorders of body systems** (p. 50)

Disorders of body systems cause predictable symptoms that may be successfully diagnosed and treated.

2.8 FIRST NATIONS SCIENCE CONTEXTS: **First Nations Australians' knowledge of body systems** (p. 52)

First Nations Australians have a complex understanding of the internal systems of animals.

2.9 SCIENCE AS A HUMAN ENDEAVOUR: **Researching diseases and treatments using organoids** (p. 54)

The use of organoids is recognised as an important way for researchers to study human diseases and their treatments.

2.10 SCIENCE INVESTIGATIONS: **Evaluating an investigation** (p. 55)

1 Does exercise affect your ability to hold your breath?
2 The structure and function of the mouth

9780170463027

Shutterstock.com/cunaplus

▲ **FIGURE 2.0.1** A failure in human body systems can result in serious illness, sometimes death.

Describe what you see in the photo above. How do you think he feels? How do you think the photo relates to the topic of body systems? How does it relate to human health?

Each year, thousands of Australians die of a disorder that could have been prevented, either by lifestyle choices or by early interventions.

▶ **What information do we need and what actions can we take to solve this problem?**

#2 SCIENCE CHALLENGE ACCEPTED!

At the end of this chapter, you can complete Big Science Challenge Project #2. You can use the information you learn in this chapter to complete the project.

Assessments
- Prior knowledge quiz
- Chapter review questions
- End-of-chapter test
- Portfolio assessment task: Project

Videos
- Science skills in a minute: Accurate and valid results **(2.10)**
- Video activities: The large intestine **(2.3)**; How does COVID affect the lungs? **(2.4)**; The heart **(2.5)**; The kidneys **(2.6)**; Heart disease **(2.7)**

Science skills resources
- Science skills in practice: Evaluating an investigation **(2.10)**
- Extra science investigations: Villi under the microscope **(2.4)**; Heart dissection **(2.5)**; Kidney dissection **(2.6)**

Interactive resources
- Crossword: The heart **(2.5)**
- Label: Parts of the digestive system **(2.3)**
- Drag and drop: Cellular respiration **(2.1)**; Chemical groups in food **(2.3)**

Nelson MindTap

To access these resources and many more, visit:
cengage.com.au/nelsonmindtap

BY THE END OF THIS MODULE, YOU WILL BE ABLE TO:

✓ describe cellular respiration

✓ state the requirements for the survival of animals.

Interactive resource
Drag and drop:
Cellular respiration

GET THINKING

Did you know that the average person eats 30 tonnes of food during their lifetime? Why do we require all that food? Does it matter what type of food we eat?

Animals need water

On average, 70 per cent of an animal's body is water. Water is an essential part of the cytoplasm of cells, and it is here that the chemical reactions of life take place. Water is also important for transporting many dissolved materials around the body. Animals lose water when sweating, urinating and breathing, so they must replace this water through their diet.

Energy comes from food

waste
unwanted or unusable material, substances or by-products

mineral
an inorganic substance present in food that is required by our body to develop and function properly

vitamin
a substance that is essential in small amounts for normal growth and activity

Energy is needed for animals to stay alive, to think and to be active. Even when we sleep, our body needs energy to make our hearts beat, to remove **waste** and to power our breathing. This energy comes from the food we eat in a process called cellular respiration (introduced in Chapter 1). As we have learned, all living things undergo cellular respiration. The inputs for cellular respiration are glucose and oxygen and the outputs are carbon dioxide, water and energy, summarised as:

glucose + oxygen → water + carbon dioxide + energy

Food provides nutrients for growth and repair

As an organism grows, food provides the substances needed to make new cells and tissues (Figure 2.1.1). Some cells, such as red blood cells, have a limited life span. Others, such as skin cells, are lost in everyday wear and tear. The routine replacement of cells also requires food, as does the ongoing repair of injuries. Sometimes, specific **minerals** must be available from food to enable cell replacement and repair. For example, the haemoglobin in red blood cells contains iron; calcium and phosphorus are required to build strong bones; and sodium is essential for nerve and muscle function.

Vitamins are also an essential requirement in the diet and are needed in small quantities. Although vitamins cannot be built into cell structures or used as a source of energy, vitamins play an important role in essential chemical reactions. There are diseases caused by a lack

Shutterstock.com/Elena Eryomenko

▲ **FIGURE 2.1.1** A variety of foods helps to provide the essential nutrients for good health.

of vitamins. Scurvy, due to a lack of vitamin C, causes bleeding in the mouth and under the skin, tiredness and generalised pain. Untreated, scurvy can lead to organ failure and death. A deficiency of vitamin A causes poor vision, reduces resistance to disease and can lead to blindness. A balanced diet of a variety of food types usually contains all the vitamins and minerals required for good health.

What is in fruit juice?

In this activity you will use information on food labels to compare the amount of energy, sugar and vitamin C in different brands of fruit juice.

You need
- at least three or four different types of fruit juice, in their original containers

What to do
1 Predict the healthiest drink from the fruit juice selections.
2 Look at the food labels and consider the amount of energy, sugar and vitamin C in the different drinks. Use a table like the one below to compare and contrast these elements.

Brand and type of juice	Energy (kilojoules/ 100 mL)	Sugar (per 100 mL)	Vitamin C (per 100 mL)

▲ FIGURE 2.1.2   Cartons of fruit juice

What do you think?
1 Can you rank the drinks from most nutritious to least nutritious?
2 Are the results what you expected?
3 How does the amount of sugar relate to the amount of energy on the food label?
4 Why is vitamin C important in the human diet?
5 Do you think fruit juices are a healthy drink? Why or why not?

2.1 LEARNING CHECK

1 **Describe** two important roles of water in animals.
2 **State** three ways in which animals lose water.
3 **State** two important functions of food in animals.
4 **List** the substances involved in cellular respiration according to their inputs and outputs.
5 **Name** two minerals and **describe** their importance in the human diet.
6 **Explain** how very small amounts of vitamins play a vital role in body health.
7 **Explain** why it is important to have a balanced diet.

BY THE END OF THIS MODULE, YOU WILL BE ABLE TO:
- ✓ describe the relationship between cells, tissues, organs, systems and organisms
- ✓ explain why multicellular organisms have multiple systems that work together.

Quiz
Levels of organisation

GET THINKING

Skim read this module, paying particular attention to the headings. Can you describe body organisation? Now read the module in detail. Were you correct?

Too big for diffusion

In Chapter 1 you learned that cells rely on diffusion to obtain their requirements and get rid of waste. This is an ideal arrangement for unicellular organisms because they are small and their surface area (cell membrane) is exposed to the outside environment. Oxygen and nutrients easily diffuse across the cell membrane from the outside environment. Waste, such as carbon dioxide, easily diffuses out of the cell into the environment (Figure 2.2.1).

Multicellular organisms are made up of hundreds, millions and sometimes trillions of cells. Imagine a big ball of one trillion cells. How would the cell in the middle of the ball obtain oxygen or food? If it relied on diffusion alone, it would die. The only way for the cell in the middle of the ball to survive is if oxygen and food are brought to it and waste is removed. This is what your respiratory, circulatory, digestive and urinary systems do. These body systems use a network of pipes and tubes to connect all the cells. The pipes and tubes bring all body cells their requirements and remove their waste (Figure 2.2.2).

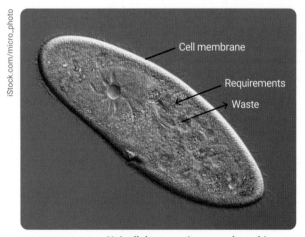

▲ **FIGURE 2.2.1** Unicellular organisms, such as this *Paramecium*, can easily obtain their requirements and get rid of waste by diffusion across the cell membrane.

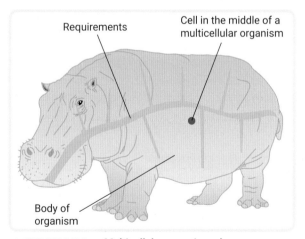

▲ **FIGURE 2.2.2** Multicellular organisms have a system of pipes and tubes to provide all cells with their requirements and to remove their waste.

Levels of organisation: cells, tissues, organs and systems

tissue
a collection of cells that have similar structures and functions

organ
a collection of different tissues that combine to perform a specific function

Cells are the basic unit of all living things. A collection of cells that perform a similar function is called a **tissue**. For example, your skin cells form skin tissue. An **organ** consists of different tissues that carry out a particular function. For example, your kidneys are organs made up of a variety of different sorts of tissues, such as muscle and

9780170463027

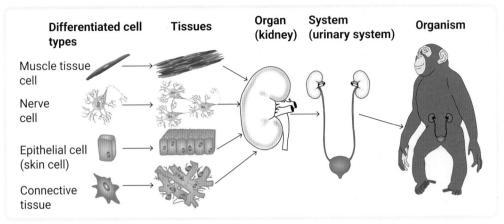

Muscle tissue cell

Nerve cell

Epithelial cell (skin cell)

Connective tissue

Differentiated cell types

Tissues

Organ (kidney)

System (urinary system)

Organism

▲ FIGURE 2.2.3   Cells make up tissues; tissues make up organs; organs make up body systems.

connective tissues. These tissues act together to achieve a common goal. When multiple organs are grouped together to carry out a specific function, this is called a **system** (Figure 2.2.3). Table 2.2.1 summarises the major organ systems of the human body.

system
a group of organs that work together to perform a specific function

▼ TABLE 2.2.1 Human organ systems

Organ system	Major organs	Major functions
Digestive	• Mouth • Oesophagus • Stomach • Intestines • Liver • Pancreas	• Physical and chemical breakdown of food • Absorption of nutrients
Respiratory	• Lungs • Trachea • Bronchi • Bronchioles	• Gas exchange
Cardiovascular	• Heart • Blood vessels (arteries, veins)	• Transportation of nutrients, gases and waste • Defence against infection
Urinary	• Kidneys • Bladder • Ureter • Urethra	• Removal of liquid waste • Water balance
Musculoskeletal	• Bones • Muscles	• Movement and support of body parts
Nervous	• Brain • Spinal cord	• Conduct messages around the body • Control of body activities

Each organ system contributes to the survival of the whole organism. Body systems work together and depend on each other. For example, the respiratory system rapidly delivers oxygen from the air to the circulatory system, which then transports the oxygen to cells in the body.

2.2 LEARNING CHECK

1 **Define** 'tissue'.

2 **Describe** the relationship between cells, tissues, organs and body systems.

3 **List** the key organs in the digestive and cardiovascular systems.

4 Using an example, **explain** the purpose of your body systems.

5 **Compare** how unicellular and multicellular organisms obtain their requirements.

Video activity
The large intestine

Interactive resources
Label: Parts of the digestive system
Drag and drop: Chemical groups in food

carbohydrates
a complex food group found in starchy foods such as bread and rice

proteins
a complex food group found in foods such as meat, fish, soybeans and cheese

fats
a complex food group found in foods such as butter and cream

mechanical digestion
the physical breakdown of food into smaller pieces

chemical digestion
the chemical breakdown of food into simpler substances

ingested
taken in; eaten

GET THINKING

Does it surprise you that the food you eat takes more than 24 hours to pass through your digestive system? In this module you will find out why it takes so long.

Digestive processing

As we have learned, food provides our body cells with energy and the raw materials for growth and repair. The foods we eat are made up of three main chemical groups: **carbohydrates**, **proteins** and **fats** (Figure 2.3.1). These substances are complex. The body cannot use food in this complex form. The role of the digestive system is to break down these foods into simpler, more useful substances, as shown in Table 2.3.1.

▼ **TABLE 2.3.1** Complex forms of substances and the simpler forms they are broken down to by the digestive system

Complex form of substance	Simple form of substance	What foods is this found in?
Carbohydrates	Simple sugars, such as glucose	Pasta, rice, bread, potatoes
Protein	Amino acids	Meat, cheese, fish, soybeans
Fats and lipids	Fatty acids and glycerol	Butter, oils

The breakdown of food can be classified as either chemical or mechanical digestion. **Mechanical digestion** (physical digestion) is when large pieces of food are broken down into smaller pieces of food. This increases the surface area of food so it can be acted on by enzymes in chemical digestion. **Chemical digestion** occurs when enzymes break down complex substances into simpler chemical forms.

Mouth

The mouth is the structure into which food is **ingested**. It contains the teeth, the tongue and salivary glands – all essential for the initial digestion of food. The teeth break food into small pieces in the process of mechanical digestion. The teeth are strong and hard, so they can chew different types of food. Smaller pieces of food have a larger total surface area for enzymes to act on during chemical digestion.

When you are eating, you may have noticed that the food becomes slippery and slimy. This is due to the addition of saliva. Saliva, made by the cells of the

▲ **FIGURE 2.3.1** This meal contains carbohydrates in the bun, protein in the meat and fat in the chips.

Shutterstock.com/Brent Hofacker

salivary glands, lubricates the food so it moves easily down the throat. Saliva also contains the enzyme **amylase**, which starts the chemical digestion of the carbohydrate **starch** into simple sugars. Your muscular tongue then moves the ball of food (**bolus**) to the back of the mouth to the entry of the oesophagus.

2.3

amylase
the enzyme that digests carbohydrates

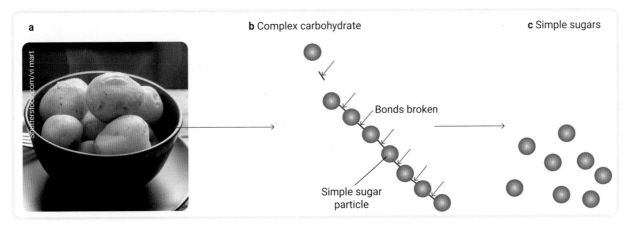

b Complex carbohydrate

c Simple sugars

Bonds broken

Simple sugar particle

▲ **FIGURE 2.3.2**   **(a)** Our teeth cut cooked potato into smaller pieces. **(b)** Enzymes start to break down the complex carbohydrates in the potato. **(c)** Simple sugar particles are released.

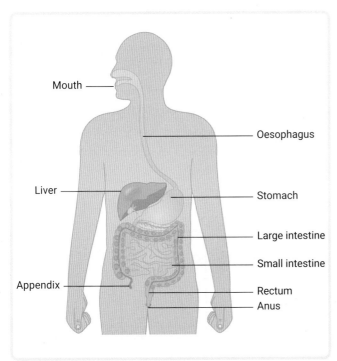

Mouth

Oesophagus

Liver

Stomach

Large intestine

Small intestine

Appendix

Rectum

Anus

▲ **FIGURE 2.3.3** The organs of the digestive system

The oesophagus

The oesophagus (Figure 2.3.3) is a long, muscular tube that transfers food from the mouth to the stomach. The circular muscles in the oesophagus contract and relax in waves so that food is pushed along. This is known as **peristalsis**. At the end of the oesophagus is a special ring-like muscle, or **sphincter**, that opens and closes the entrance to the stomach. This sphincter prevents the acidic contents of the stomach from flowing back up into the oesophagus and irritating the oesophagus wall – a condition known as **heartburn**.

starch
a complex carbohydrate found in potatoes and other plants; also a form of glucose storage in plants

bolus
a ball of food that passes into the oesophagus from the mouth

peristalsis
a progressive wave of contraction and relaxation of the digestive tract

sphincter
a ring of muscle that can close off a tube

heartburn
a burning feeling in the oesophagus caused by rising stomach acid

hydrochloric acid
a type of acid; in the stomach it helps to digest food

chyme
partially digested food that passes from the stomach to the small intestine

The stomach

The stomach (Figure 2.3.4b and c) is a muscular sac at the end of the oesophagus. Its wall contains many folds that allow it to expand to hold up to 3 litres after a large meal. This is 60 times its usual volume. The stomach wall contains glands that secrete protein-digesting enzymes and **hydrochloric acid** that make the stomach very acidic. The mixture of food, acid and enzymes formed in the stomach is called **chyme**. The muscular wall of the stomach contracts to mix and churn the chyme, another example of mechanical digestion.

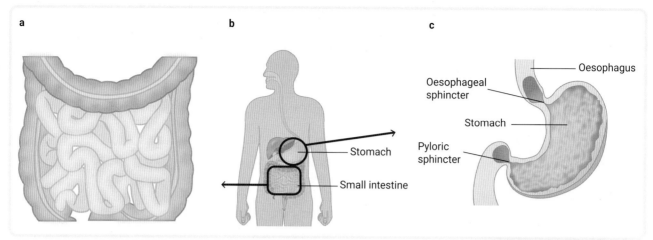

▲ FIGURE 2.3.4 **(a)** The small intestine is a long, tightly coiled tube, in which protein, fat and carbohydrates are digested and their products absorbed. **(b)** The location of the small intestine and stomach. **(c)** A cutaway section of the stomach, showing its sphincters

The small intestine

Chyme is released in small amounts from the stomach into the small intestine through another sphincter. Laid out straight, the small intestine (Figure 2.3.4a and b) would stretch to more than 6 metres. Peristalsis continues in the small intestine. This helps to mix the chyme with enzymes that digest protein, fat and carbohydrates.

villi
small finger-like projections on the cells of the small intestine that increase surface area

Magnification of the wall of the small intestine (Figure 2.3.5) shows that it is made up of **villi** (singular: villus). These finger-like extensions increase the surface area of the small intestine and take blood vessels close to the chyme. This arrangement ensures that the products of digestion, including amino acids, fatty acids and glucose, are rapidly absorbed into the blood along with large amounts of water. Any material that has not yet been absorbed remains in the small intestine and continues along the digestive tract to the large intestine.

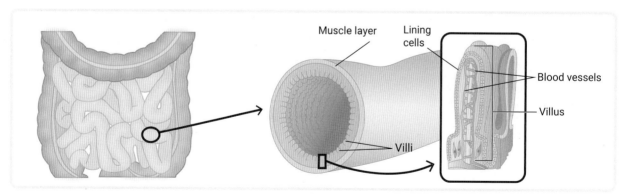

▲ FIGURE 2.3.5 A magnified view of the wall of the small intestine lined with villi, which increase the surface area that is available for absorption of nutrients

9780170463027

The large intestine

The small intestine attaches to the large intestine (Figure 2.3.6) near the **appendix**. At only 1.5 metres long, the large intestine is named for its width rather than its length. The main function of the large intestine is to absorb water, vitamins and minerals into the capillaries of the circulatory system.

The cellulose cell walls of plants cannot be broken down by the human digestive system. This is commonly referred to as **fibre**, or 'roughage', in the diet and is beneficial in two ways. Fibre gives bulk to the intestinal contents and speeds up the passage of food through the digestive tract, preventing constipation. Bacteria living on the fibre in the large intestine produce some vitamins as well as gases.

appendix
a small tube-shaped sac attached to, and opening into, the lower end of the large intestine

fibre
the indigestible parts of plants

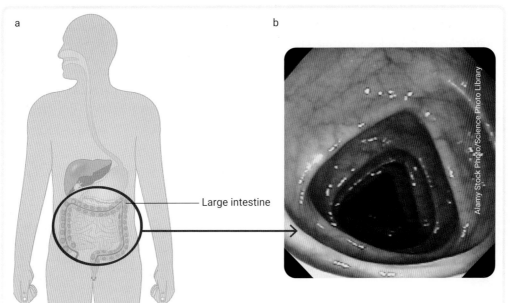

a

b

Alamy Stock Photo/Science Photo Library

Large intestine

faeces
undigested waste material

anus
the external opening of the rectum, through which faeces leave the body

bowel motion
the process of egesting faeces through the anus

egest
to pass out of the body

▲ **FIGURE 2.3.6** **(a)** The location of the large intestine, surrounding the small intestine. **(b)** The inside of the large intestine has many blood vessels.

The rectum and the anus

At the end of the large intestine is the rectum (Figure 2.3.7). The rectum is a storage facility that can stretch to hold undigested material, called **faeces**. At the end of the digestive tract is the **anus**, through which faeces leave the body. When a **bowel motion** occurs, the anal sphincter – which is a specialised muscle that is normally closed – opens to allow waste to be **egested**.

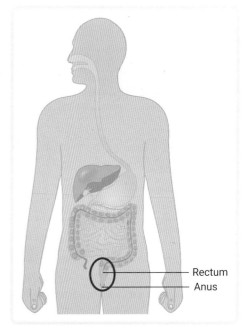

Rectum
Anus

▲ **FIGURE 2.3.7** The location of the rectum and the anus

Chapter 2 | Animal systems **39**

How do we compare?

How do the digestive systems of other animals differ from ours? Most people have an **omnivorous** diet that includes food sourced from both plants and animals. The human digestive system is suited to this diet. The digestive systems of other animals are also suited to their own diets.

Koalas, for example, have a **herbivorous** diet that is restricted to the leaves of certain eucalyptus trees. These leaves are very tough and difficult to digest and are low in protein and energy. They also contain poisons. Koalas have sharp front teeth to clip leaves. Their back teeth are flat and wide for grinding the leaves (Figure 2.3.8a). The chewed material passes down the koala's oesophagus to the stomach and then into the small intestine. Between the small and large intestines is the caecum. The **caecum** is a pouch or large tube-like structure at the beginning of the large intestine. It receives undigested food material from the small intestine.

Humans also have a caecum, but it is underdeveloped. In koalas, the caecum is 2 metres long (Figure 2.3.9). It contains cellulose-digesting micro-organisms that break down the cell walls of the eucalypt leaves. This releases the nutrients within the cells for use by the koala. It also neutralises the poisons.

omnivorous
describes an organism that feeds on both plants and animals

herbivorous
describes an organism that feeds on plants only

caecum
a pouch or large tube-like structure at the beginning of the large intestine; receives undigested food material from the small intestine

carnivorous
describes an organism that feeds on animals only

a

b

First Nations Australians brought dingoes to Australia about 4000 years ago. Dingoes have a **carnivorous** diet that includes wombats, rabbits, possums, wallabies and kangaroos. The teeth of a dingo are those of a hunter. Their incisors at the front of the jaw are used for nibbling and are especially useful for stripping meat when eating. Their curved canines are larger and stronger than the incisors, and are used for holding objects in their mouths and tearing meat. Molars are positioned towards the back of the jaw and are used for crushing and grinding (Figure 2.3.8b).

Dingoes have a long small intestine (Figure 2.3.9), which is important for digesting and absorbing the fat and protein in meat. Unlike koalas, they have a very small caecum and large intestine because they eat very little plant material containing cellulose.

▲ **FIGURE 2.3.8** **(a)** A koala's teeth are suited to its diet of eucalyptus leaves. **(b)** A dingo's teeth are suited to its diet of meat.

9780170463027

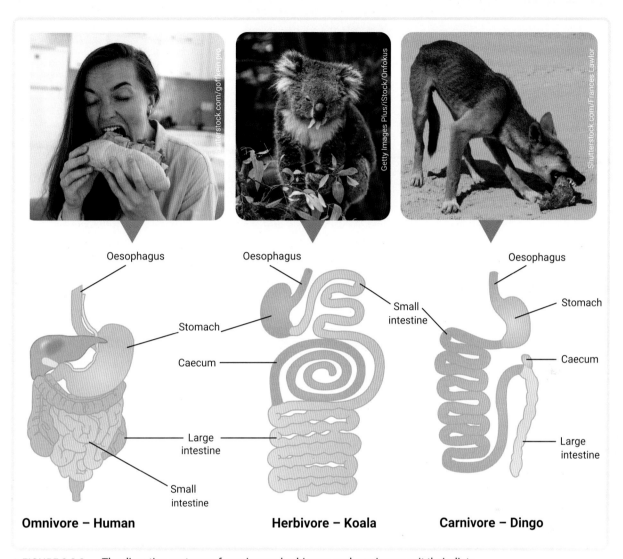

Omnivore – Human   Herbivore – Koala   Carnivore – Dingo

▲ FIGURE 2.3.9   The digestive systems of omnivores, herbivores and carnivores suit their diets.

2.3 LEARNING CHECK

1 What is the role of the digestive system?

2 What are enzymes? What is their role in digestion?

3 **Explain** the difference between mechanical digestion and chemical digestion. Give an example of each from the human digestive system.

4 **Explain** why it is important for our body's digestive system to break down complex foods into their simpler forms.

5 Where is water absorbed along the digestive tract?

6 What is the role of a sphincter? Is its structure suited to its function? **Explain** your answer.

7 **Construct** a table.

  a In the first column of the table, place the following organs in their correct order from mouth to anus: mouth, large intestine, stomach, rectum, oesophagus, caecum, small intestine, anus.

  b In the second column of the table, **describe** the function of each of these organs.

BY THE END OF THIS MODULE YOU WILL BE ABLE TO:

✓ explain how the structures of cells and organs are related to their functions in the respiratory system.

Video activity
How does COVID affect the lungs?

Extra science investigation
Villi under the microscope

inspiration
breathing in

expiration
breathing out

diaphragm
a sheet of muscle under the lungs that assists with inhalation and exhalation

trachea
a tube that runs from the back of the throat to the bronchi

bronchi (singular: bronchus)
tubes that branch off the trachea to the left and right lung

cartilage
flexible tissue that makes up part of the skeleton

GET THINKING

Skim read the module, paying particular attention to the figures. Does the information in the figures fit with what you already know about respiration?

As you saw in Module 2.1, when glucose is broken down during cellular respiration, oxygen is consumed and carbon dioxide and water are produced. The role of the respiratory system is to bring oxygen into the body and to remove the waste carbon dioxide.

Breathing

Air enters the body through the mouth and nasal passages and moves into the lungs. This process is called **inspiration**, or breathing in. As the air enters the body, it is warmed and moistened. On **expiration**, or breathing out, air is forced out of the lungs. Muscles in the chest and the **diaphragm** are responsible for increasing and decreasing the size of the chest cavity so that air is pushed in and out (Figure 2.4.1).

During inspiration, air passes into the windpipe, or **trachea**. The trachea is a long tube running from the back of the nasal passage to the **bronchi (singular: bronchus)**. Rings of **cartilage** around the trachea and bronchi provide support to keep these tubes open. Each bronchus, formed by the branching of the trachea, takes the air deep into each lung.

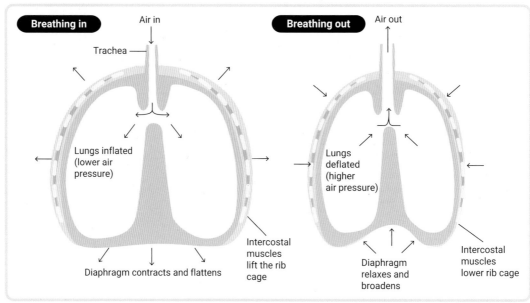

▲ **FIGURE 2.4.1** Breathing in and out involves muscles lifting and lowering the rib cage.

Gas exchange

The bronchi branch into **bronchioles**, which are smaller tubes that continue branching into even smaller tubes until they end in **alveoli (singular: alveolus)** (Figure 2.4.2). The alveoli are very small sac-like structures surrounded by very small blood vessels called capillaries (Figure 2.4.3). The alveoli have a moist surface to help with oxygen uptake. The oxygen in the air within the alveoli dissolves into the moist surface. This allows the oxygen to diffuse across the alveoli and capillary surfaces and enter the bloodstream. Waste carbon dioxide in the bloodstream moves the opposite way, from the blood to the air inside the alveoli. From here, carbon dioxide travels back up the bronchioles, bronchi and trachea and is breathed out when you exhale.

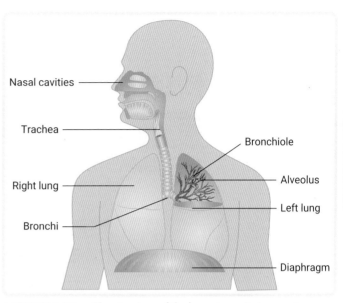

▲ FIGURE 2.4.2   The structure of the human respiratory system

The shape and number of the alveoli provide a huge surface area for gas exchange. If the whole surface of the alveoli in a human was laid out flat, it would cover approximately 140 square metres: the size of a tennis court. This large surface area allows enough oxygen into – and carbon dioxide out of – your body for your cells to continually carry out cellular respiration.

Inspired and expired air contain different percentages of oxygen and carbon dioxide, as shown in Table 2.4.1.

▼ TABLE 2.4.1 The approximate composition of inhaled and exhaled air

	Nitrogen (%)	Oxygen (%)	Carbon dioxide (%)
Inspired air	78	21	0.04
Expired air	78	14	4.40

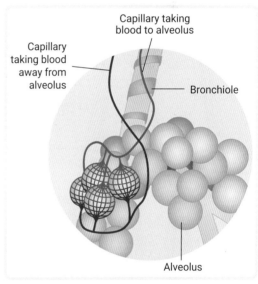

▲ FIGURE 2.4.3   An alveolus, showing the close connection to the circulatory system

2.4 LEARNING CHECK

1 **Describe** the journey of an oxygen particle from the air to an alveolus.

2 **Explain** why the:

  a composition of exhaled air is different from that of inhaled air.

  b percentage of nitrogen in inhaled air and exhaled air does not change.

3 Why is a large surface area important in the respiratory system?

4 **Describe** how gases move between the air inside an alveolus and the blood.

bronchiole
a smaller tube made when the bronchi divide

alveoli (singular: alveolus)
air sacs at the end of the bronchioles in the lungs

BY THE END OF THIS MODULE, YOU WILL BE ABLE TO:

✓ explain how the structures of cells and organs are related to their functions in the cardiovascular system.

Video activity
The heart

Interactive resource
Crossword: The heart

Extra science investigation
Heart dissection

GET THINKING

Look at the glossary terms in the margins in this module. Are you familiar with some of the terms? Are there any terms that you would like to know more about?

The human cardiovascular system is made up of the heart, blood vessels and blood (Figure 2.5.1). The cardiovascular system links all of the systems of the human body. It transports oxygen from the lungs and the products of digestion directly to cells. It also takes away waste from cells and transports it to the lungs and kidneys for removal from the body.

The heart

The heart is a muscular pump about the size of your fist. It is located between your lungs, slightly to the left of centre. The heart has a natural pacemaker that keeps it pumping regularly. Table 2.5.1 shows that heavy exercise can raise the heart rate and increase the volume of blood pumped through it.

▼ TABLE 2.5.1 Typical heart rate and blood volume pumped at rest and during heavy exercise

	Heart rate (beats/min)	Volume of blood pumped out for each ventricular contraction (mL)	Volume (L/min)
At rest	70	70	4.9
Heavy exercise	180	160	28.8

Figure 2.5.2 shows the structure of the heart and the blood vessels entering and leaving the heart. The heart is divided into four sections, or **chambers**. The two chambers at the top of the heart are the **atria (singular: atrium)**. The two chambers at the bottom of the heart are the **ventricles**. The walls of the atria and ventricles are different thicknesses. The atria have thin muscular outer walls compared with the thick muscular walls of the ventricles. The ventricle chambers require thicker walls because they need to pump blood much further and with more force than the atria.

Blood circulation

It is usual to refer to the left and right sides of your heart as they are located in your body. That is, the right side of the heart is on the right side of your body and the left side is on the left of your body.

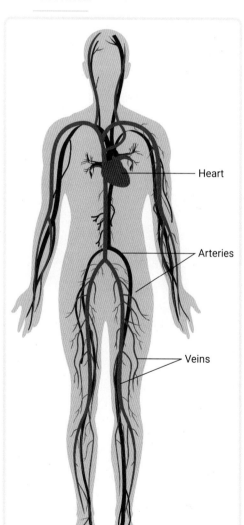

— Heart

— Arteries

— Veins

▲ FIGURE 2.5.1 The human cardiovascular system. Veins (coloured blue) transport blood to the heart, and arteries (coloured red) transport blood from the heart.

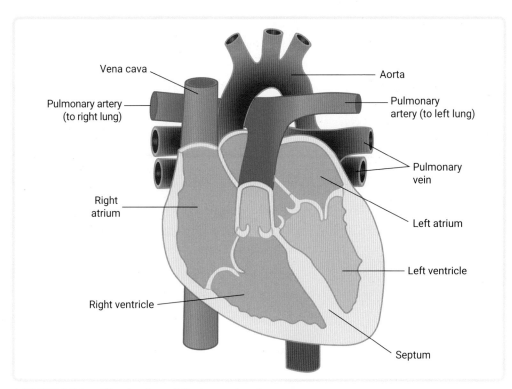

▲ **FIGURE 2.5.2**   The structure of the human heart

chamber
one of the compartments that form the structure of the heart; there are four chambers in the human heart

atria (singular: atrium)
the two upper chambers in the human heart, which receive blood from veins

ventricles
the two lower chambers in the human heart, which pump blood to either the lungs or the rest of the body

vena cava
a large vein in humans that brings blood to the heart from all parts of the body

pulmonary artery
a blood vessel in humans that takes blood from the heart to the lungs

pulmonary vein
a blood vessel in humans that returns blood from the lungs to the heart

aorta
a large artery in humans that takes blood from the left side of the heart to the body

septum
the dividing wall between the left and right sides of the human heart

Blood from the body enters the right atrium from two large, thin-walled vessels called the **vena cava**. This blood is low in oxygen (coloured blue in Figure 2.5.1) because the oxygen has been used up by the cells in the body. It also contains high levels of carbon dioxide. The blood is under low pressure because much of it has moved up from below your heart. Once in the heart, the blood moves from the right atrium, down through a one-way valve, into the right ventricle. From here, the blood is pumped through another one-way valve into the **pulmonary artery** to travel to the lungs.

In the lungs, waste carbon dioxide is removed from the blood and oxygen is added to it. Oxygenated blood returns from the lungs in the **pulmonary vein** and enters the left atrium. Blood then moves through another one-way valve into the left ventricle. When the ventricle contracts, blood is pumped through yet another valve into the **aorta**, to begin its journey around the body. In Figures 2.5.2 and 2.5.3, you will notice that the wall of the left ventricle is thick and muscular. This ensures that blood can be pumped with enough force to travel all the way around the body. Blood on the right side of the heart never mixes with blood on the left side of the heart. A wall down the middle of the heart keeps both sides separate. This wall is called the **septum**.

There are four valves in the human heart; one at the exit of each chamber. Heart valves are made of flaps of tissue that only open in one direction. If blood tries to flow backwards, the valve will snap shut, preventing back flow.

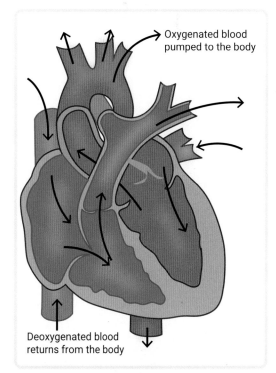

▲ **FIGURE 2.5.3**   The circulation of blood through the left and right sides of the heart

Blood vessels

The major types of blood vessels are (Figure 2.5.4):

- **arteries** – the vessels that carry blood *away from* the heart
- **veins** – the vessels that carry blood *to* the heart
- **capillaries** – the very small vessels that intertwine with your cells. It is here that oxygen and carbon dioxide move between blood and tissues.

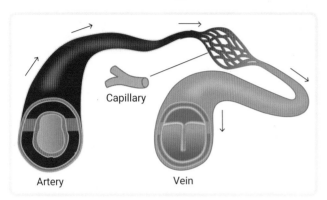

▲ **FIGURE 2.5.4**   The major types of human blood vessels

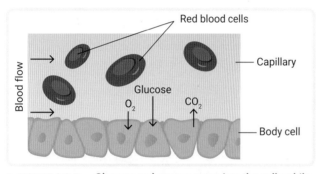

▲ **FIGURE 2.5.5**   Glucose and oxygen move into the cells while carbon dioxide moves out.

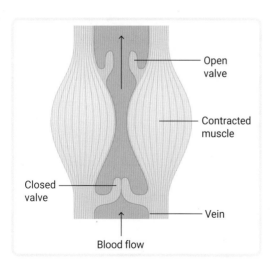

▲ **FIGURE 2.5.6**   The valves in a vein control the direction of blood flow.

Arteries and veins have different structures to accommodate the different pressures of blood flowing through them. Arteries have thick, muscular walls to withstand the high pressure used to pump blood away from the heart.

Blood is pumped from the left side of the heart into the aorta – a very large, thick-walled artery. From here, the blood moves into smaller and smaller arteries, finally entering capillaries. Capillaries are found in every tissue of the body. Materials are exchanged within capillaries. Glucose and oxygen move into the cells to be used in cellular respiration, while waste carbon dioxide moves out of the cells into the blood (Figure 2.5.5).

Blood from tissues travels in the veins back to the heart. As veins contain blood at lower pressure, they have thin walls. These walls can stretch and sometimes blood pools in them, rather than continuing to flow through the vessel. Veins contain one-way **valves** to stop blood from the lower body moving in the wrong direction (Figure 2.5.6). A valve is a device that ensures that substances flow in one direction. Contracting and relaxing leg muscles also helps to push blood along the veins back to the heart.

Blood

Blood is the transport fluid that links all of your organs. It carries oxygen and nutrients to every part of your body and transports waste away. There are approximately 5 litres of blood in your body, confined to blood vessels, unless they are damaged by injury.

Blood is made up of various components (Table 2.5.2). **Plasma** is the straw-coloured watery component. It contains glucose, proteins, **hormones** and carbon dioxide. Hormones are chemical messengers that control and regulate certain cells and tissues. **Red blood cells** are also in the plasma. They contain **haemoglobin** that

9780170463027

attaches to oxygen to transport it around the body. **Platelets** are cell fragments that assist in clotting blood in a wound. **White blood cells** are mostly involved with fighting infection as part of the **immune system**. You will learn about the immune system in Year 9.

▼ TABLE 2.5.2 The components of human blood

Constituent	Appearance	Features	Function
Plasma		Makes up about 55% of the volume of blood Consists of 90% water with nutrients, including proteins, glucose, minerals, hormones and carbon dioxide	A body fluid Carries blood cells and other substances around the body
Red blood cells		Make up 40–45% of the volume of blood Have no nucleus and a limited life span of 3–4 months Made in the bone marrow	Carry oxygen bound to haemoglobin in their cells
White blood cells		Make up about 1% of the volume of blood Made in the bone marrow	Defend the body against disease
Platelets		Make up less than 1% of the volume of blood Cell fragments Survive about a week	Blood clotting

artery
a blood vessel in humans that carries blood away from the heart

vein
a blood vessel in humans that carries blood to the heart

capillary
a very small blood vessel in humans, located in between the smallest arteries and smallest veins

valve
a structure in the human heart and veins that prevents backflow of blood

plasma
the watery component of human blood, in which blood cells are suspended

hormone
a chemical messenger

red blood cell
a blood cell that carries oxygen

haemoglobin
the component of red blood cells that binds with oxygen

platelets
fragments of cells that act in blood clotting

white blood cell
a blood cell that is part of the human immune system

immune system
a complex system that defends the human body against infection and disease

2.5 LEARNING CHECK

1 What three components make up the cardiovascular system?
2 **Outline** the role of the blood in supplying your body with its requirements and removing waste.
3 If a person had more white blood cells than normal in their blood, what conclusion could you come to?
4 Refer to Figure 2.5.7.
   a **Name** the structures labelled A to F.
   b Trace the path of blood through the heart by listing, in the correct order, the structures in the diagram starting from F.
5 What would happen if one of the valves in your heart became a two-way valve?
6 **Explain** the benefit of an increase in heart rate during heavy exercise.

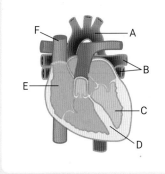

▲ FIGURE 2.5.7

BY THE END OF THIS MODULE, YOU WILL BE ABLE TO:

✓ explain how the structures of cells and organs are related to their functions in the urinary system.

Video activity
The kidneys

Extra science investigation
Kidney dissection

excretion
the process of eliminating or expelling waste matter

urea
nitrogenous waste that is created as amino acids are broken down in the human body

kidneys
the excretory organs of mammals

GET THINKING

Imagine that you are marooned at sea in a lifeboat a long way from shore. There is no fresh water available. Should you drink sea water? Could you excrete that much salt?

Waste removal

The human body is not 100 per cent efficient at converting raw materials into useful substances or energy. Unwanted materials, called waste, are removed from the body in the process of **excretion**. One such waste is **urea**, a toxic substance produced in the breakdown of excess amino acids. Other materials that are excreted from the body include water and salts that have been taken in from food.

In mammals, the organs that excrete waste are the **kidneys**, lungs and skin. You already know that the lungs excrete carbon dioxide during exhalation. Both the skin and the kidney excrete urea, water and salts, including sodium chloride. Excretion by the skin takes place during sweating (Table 2.6.1).

▼ TABLE 2.6.1 Typical values for water intake and output in humans

The ways water enters the body	Volume (mL)	The ways water leaves the body	Volume (mL)
Drink	1400	Urine	1500
Food	800	Sweat	450
From cellular respiration	300	Breath	450
		Faeces	100
Total	2500	Total	2500

This module will focus on the role of the urinary system in excretion. As well as excreting waste, the kidneys also maintain the delicate balance between the amount of water in the blood and the concentration of solutes dissolved in it.

Kidney function

The kidneys are two bean-shaped organs located at the back of the upper abdomen. The renal artery is a branch of the aorta that brings oxygenated blood containing waste to the kidneys. The renal vein takes deoxygenated, filtered, clean blood from the kidneys back into general circulation (Figure 2.6.1).

nephron
the structure in the human kidney where filtration of the blood occurs

filtration
the process in the kidney where all materials, except for protein and blood cells, are forced out of the bloodstream

Formation of urine

Waste is removed from the blood by tiny structures in the kidneys called **nephrons**. During **filtration**, the blood capillaries and tubules of the nephrons are responsible for

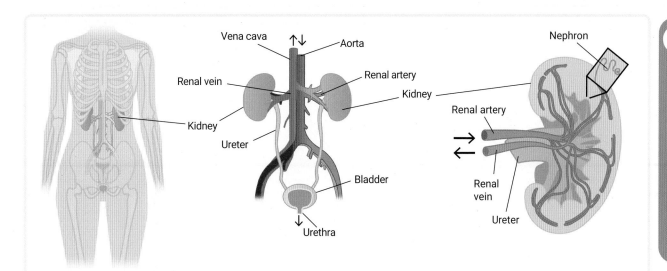

▲ FIGURE 2.6.1   The human excretory system

removing both waste and useful material from the blood. Then, as these materials move through the nephron, blood reclaims the useful substances, as well as some water and solutes from the tubules.

Waste, such as urea, is not reabsorbed into the blood. Instead, the urea leaves the kidney dissolved in **urine**. A tube, called the **ureter**, carries urine from each kidney to the bladder. The bladder stores urine until it is removed from the body via a single tube called the **urethra** (Figure 2.6.1).

The amount of water consumed in food and drink can vary. The body, via the kidneys, adjusts for excess water intake by increasing urine output. Conversely, it adjusts for increased exercise or decreased water intake by reducing urine. Therefore, the kidneys not only prevent the build-up of waste in the body, they also help maintain water balance by controlling the volume, composition and pressure of body fluids.

urine
liquid containing multiple waste products, especially urea

ureter
the tube that carries urine from the human kidney to the bladder

urethra
the tube that carries urine from the bladder to the outside of the human body

2.6 LEARNING CHECK

1 **List** wastes that need to be excreted from the body and name the organs that remove them.
2 **State** the functions of the:
   a kidney.
   b urethra.
   c bladder.
3 **Describe** the role of a nephron.
4 **Predict** the effect of each of the following on the amount and concentration of urine.
   a Playing tennis
   b Having a bath
   c Drinking a large volume of soft drink
5 Kidneys only remove waste. **Explain** why this statement is not true.

BY THE END OF THIS MODULE, YOU WILL BE ABLE TO:

✓ describe the causes, symptoms and treatment of common disorders of body systems.

Video activity
Heart disease

GET THINKING

In previous modules we looked at different body systems. Recall the main functions of each of the body systems that you learned about. In this module, you will find out what can go wrong.

Symptoms and diagnosis

The first sign of a disorder is usually one or more symptoms. **Symptoms** are the effects on the body of a disorder or disease. Even if you know the symptoms, making a **diagnosis** of an internal disorder is often difficult. Doctors use an **endoscope** to look inside the body without having to perform surgery. Internal abnormalities revealed through endoscopy include inflammation, cancer, polyps, tumours, ulcers and other diseases and conditions.

Kidney stones

Kidney stones form when minerals congregate into masses of crystals in the kidney. They commonly range in size from as small as a grain of sand to as big as a golf ball. Occasionally, stones can become very large, or can travel into the bladder where they may continue to grow (Figure 2.7.1). Very small stones can be excreted in the urine. However, larger kidney stones usually become painful when they move along the ureter or the urethra. The exact cause of kidney stones is not known. The risk of forming kidney stones can be reduced by drinking plenty of water.

symptom
an indication of a disorder or disease

diagnosis
the identification of the nature of an illness

endoscope
an instrument used to look inside the human body

cholesterol
an insoluble, waxy substance

atherosclerosis
the build-up of deposits on the inner wall of arteries

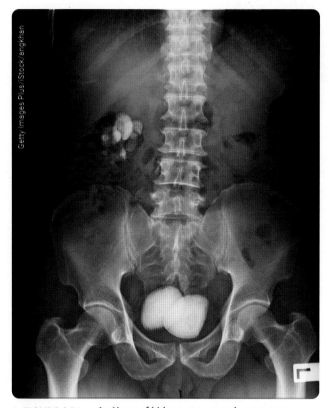

▲ **FIGURE 2.7.1** An X-ray of kidney stones and cancerous bladder stones

Coronary artery disease

Coronary artery disease occurs when the arteries that supply the heart with blood become blocked. Blockages form when **cholesterol** builds up on the walls of the coronary blood vessels, narrowing them and allowing less blood to pass through. This is called **atherosclerosis**. If the blood vessel becomes completely blocked, an area of the heart is deprived of blood supply, resulting in tissue damage. This leads to a heart attack. Symptoms of a heart attack include sudden onset of chest pain, breathlessness, nausea and cold, clammy skin. The resulting damage to the heart may be very severe, leading to heart failure and even death.

There is no one cause of coronary heart disease, but **risk factors** can increase a person's chance of developing the condition. Some factors, such as age, gender or family history, cannot be altered. Other factors, such as smoking and not getting enough exercise, are under our control. Medication can reduce risk factors such as high blood cholesterol and high blood pressure. In some cases, a patient may need surgery to improve the blood flow to the heart. A coronary artery bypass uses a length of blood vessel from the arm or leg to bypass the blockage. This will not cure the disease, but is likely to reduce symptoms and lower the chances of the person having a heart attack in the future.

risk factor
a condition or behaviour that increases the likelihood of a person developing a disease or health disorder

Peptic ulcers

Peptic ulcers are open, inflamed sores on the lining of the stomach. They are also known as gastric ulcers. Stomach pain is usually the first symptom of a peptic ulcer. Other, less common, symptoms include indigestion, heartburn, nausea, vomiting and loss of appetite. For many years, peptic ulcer disease was treated with antacids. The disease was thought to be caused by too much acid in the stomach due to stress or spicy food.

In the 1980s, two Australian doctors, Barry Marshall and Robin Warren, did research showing that peptic ulcers were actually caused by a species of bacteria, *Helicobacter pylori* (Figure 2.7.2). Dr Marshall was so determined to convince the world that bacteria – not stress or food – caused ulcers that he drank a batch of *Helicobacter pylori*. Five days later he was vomiting and had severe stomach inflammation. He quickly recovered once he started taking antibiotics. In 2005, the two doctors were awarded the Nobel Prize in Physiology or Medicine for their pioneering work on peptic ulcer disease.

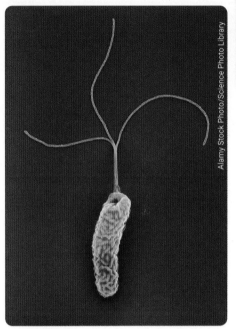

▲ **FIGURE 2.7.2** The *Helicobacter pylori* bacterium

2.7 LEARNING CHECK

1 **Define**:
   a symptom.
   b diagnosis.
2 **Describe** how an endoscope could be used to diagnose a gastric ulcer.
3 What causes a peptic ulcer and how is it treated?
4 **Predict** two symptoms a person would experience if they had a kidney stone.
5 The heart is full of blood. **Explain** why it needs its own blood supply through the coronary arteries.
6 Copy the following table and research the diseases listed.

Disease	Body system	Cause	Symptoms	Treatment
Lung cancer				
Colon cancer				
Nephritis				
Asthma				

2.8 First Nations Australians' knowledge of body systems

IN THIS MODULE, YOU WILL:
✓ explore First Nations Australians' long-held understanding of the internal systems of animals.

Understanding internal systems of humans

Prior to European colonisation, First Nations Australians obtained almost everything they needed to live from Country/Place. They faced and had to overcome many challenges to obtain essential resources, including food, water and medicines.

First Nations Australians shared their knowledge within and between related cultural groups. This includes groups who shared social, linguistic, spiritual, economic or kinship connections. Knowledge of physiology and medicine has long been shared through cultural narratives and pictorial representations over many generations. Their traditional medical knowledge, accumulated and refined over thousands of years, facilitated the treatment of ailments and injuries. First Nations Australians had, and continue to maintain, a detailed knowledge of the medicinal benefits of many plant species. They also developed and used different delivery methods; for example, through ingestion, inhalation or absorption through the skin. This shows the detailed understanding First Nations Australians have of the human body and how it works.

☆ ACTIVITY 1

1 **Relate** each delivery method to a specific internal system.
2 For one of the identified systems, **explain** how the treatment is delivered into the body.

Representing internal systems of animals

First Nations Australians have been hunting large animals such as kangaroos and emus for thousands of years, using a variety of methods including fires and nets. Unlike colonists, who primarily hunted native animals for sport or for eradication, First Nations Australians hunted and killed only what they needed.

First Nations hunters had a good understanding of kangaroo physiology. They could identify tracks, how the animal would behave if it detected movement, and where on its body was best to spear it so that the kangaroo died with little suffering. This information was gained through an ongoing cycle of formal instruction, which included demonstration, observation of people with proficient hunting skills and personal practice. It could often take years for a person to become a successful hunter.

When an animal was killed, nothing was wasted. All the edible parts were eaten, the skin was used to make cloaks and blankets, sinew was used as bindings, and teeth and bones were used as tools and body ornaments. A good understanding of an animal's internal structure is needed in order to remove and use all parts of the animal. For example, First Nations Australians would be able to locate and carefully remove highly sought-after organs such as the liver and kidneys.

▲ **FIGURE 2.8.1** An X-ray-style bark painting of two emus by Yirawala, a Kuninjku artist from western Arnhem Land, NT

▲ **FIGURE 2.8.2** An X-ray style bark painting of kangaroo and hunter from western Arnhem Land, NT; artist unknown

Examine Figures 2.8.1 and 2.8.2.

1 **Name** each of the organisms drawn.

2 **List** the body systems and organs you can identify in each of the drawings.

3 **Explain** how these drawings would help a hunter.

4 **Explain** how these drawings would help maximise use of all the animal's parts.

5 Research a scientific diagram of the internal organs of one of the organisms you named in Question 1 and **identify** similarities with the drawing.

6 **Explain** why the type of artwork shown in Figures 2.8.1 and 2.8.2 may be referred to as X-ray art.

7 Research the properties of sinew and **explain** why First Nations Australians used it as bindings.

2.9 Researching diseases and treatments using organoids

BY THE END OF THIS MODULE, YOU WILL BE ABLE TO:

✓ describe the role of organoids in the research and treatment of diseases.

What are organoids?

organoid
a small organ grown in the laboratory from stem cells

Stem cells derived from both healthy people and those with disease are routinely used to grow small organs in the lab, known as **organoids**. The use of organoids is now recognised as an important way for researchers and pharmaceutical companies to study human diseases and their treatments. Researchers have been able to produce organoids that resemble, for example, the brain, kidney, intestine, stomach, lungs and liver. The stem cells follow their own genetic instructions to differentiate and self-organise, forming tiny structures composed of many cell types that resemble miniature organs.

Organoids for research

Research using organoids will enable scientists to develop a better understanding of many diseases. This research could help develop treatments, including new drugs to target different aspects or stages of a disease. Currently, promising drug treatments are first tested on animals before they can be tested on humans. One disadvantage of testing new drugs or treatments on animals is that animal tissue may respond differently to human tissue. Testing new drugs on specific disease organoids solves this problem, making testing cheaper, faster and more effective. It also removes the ethical issue of using animals for drug screening. As organoids can be derived from a patient's own cells, it might also be possible to devise and test individualised treatment plans.

2.9 LEARNING CHECK

1 Organoids are kept alive in cell culture for months or even years. **Reflect** on the requirements of animals outlined in Module 2.1. List four conditions that would be needed to grow organoids successfully in a liquid medium.

2 Consider the use of organoids for researching diseases and treatments.

   a **Describe** at least three benefits of using organoids to test new pharmaceutical drugs and classify these benefits as economic, ethical or social.

   b Which type of benefit do you think is the most important?

3 Some people refer to organoids as artificial organs. Do you think this is an accurate name? **Explain** your answer.

SCIENCE SKILLS IN FOCUS

IN THIS MODULE, YOU WILL FOCUS ON LEARNING AND IMPROVING THESE SKILLS:

- ▶ evaluate an investigation by considering reliability and validity
- ▶ describe the effect of exercise on the ability to hold your breath
- ▶ explain the action of saliva and salivary amylase in the mouth.

When you evaluate your method and results in an investigation, you should reflect on what you did, how well it was done and the results you have obtained. Consider the following questions.

▶ **Are the results reliable?**

- Reliable means getting very similar results when repeated measurements are made. If you did the same investigation again, would you get the same results?
- Did you follow the same method exactly every time?
- How many people did you test? Repeating the test on several people increases the reliability of the data.

▶ **Are the results valid?**

- An investigation is valid if it measures what it is supposed to measure.
- Were the variables controlled properly?
- Was the same equipment always used?
- Were there any sources of error?

▶ **Can you make any suggestions for improvement?**

- How could you improve the method or better control the variables?
- Were there variables that you assumed were controlled? What were they?
- Could you collect the results differently? Provide reason(s) for your answer.

▶ **Can you suggest any further investigations?**

- Are there new questions that could be followed up or explored in another investigation?

INVESTIGATION 1: DOES EXERCISE AFFECT YOUR ABILITY TO HOLD YOUR BREATH?

BACKGROUND

The levels of carbon dioxide, and to a lesser extent oxygen, in the blood are important factors in our rate of breathing. If the carbon dioxide concentration is too high, the brain sends impulses to the muscles of the ribs and diaphragm to contract. This brings fresh air into the lungs and increases the rate at which carbon dioxide is lost in exhaled air. The overall effect is a lowering of the carbon dioxide concentration in the blood.

AIM

To investigate the effect of the duration of exercise on breath-holding ability

YOU NEED

☑ 2 stopwatches

WHAT TO DO

1. Work in groups of three. One person times and records the duration of exercise, one person times and records the duration of breath-holding and the other person is the subject. Rotate through these roles if time permits.

2. Before you start, read Table 2.10.1, which shows how to control important variables.

 Warning

Anyone with a medical issue – for example, asthma – should take the role of recorder and should not be the subject of this investigation.

Video
Science skills in a minute: Accurate and valid results

Science skills resource
Science skills in practice: Evaluating an investigation

▼ TABLE 2.10.1 The ways in which variables should be controlled throughout the investigation

Variable	How to control the variable
Intensity of exercise	Moderate jogging on the spot
Time between exercise	Wait 1 minute before doing the next bout of exercise
Depth of breath before holding the breath	Inhale gently before holding the breath
Amount of clothing	Either remove warm clothing before starting or leave it on throughout

3   Draw a table like the one below to record your results.

Duration of exercise (minutes)	Length of breath holding (seconds)			
	Subject 1	Subject 2	Subject 3	Average
0				
0.5				
1				
2				
3				
4				

4   Determine a resting (duration of exercise: 0 seconds) breath-holding ability by timing how long the subject can hold their breath while sitting down, without having exercised. Record the result. Allow the subject to rest for 1 minute.

5   The subject exercises by jogging moderately on the spot for 30 seconds (duration of exercise: 0.5 minutes). As soon as the subject stops exercising, the other timer checks how long they can hold their breath. Record the result.

6   Allow the subject to rest for 1 minute.

7   Repeat Step 5, with the subject exercising for 1 minute. Repeat for exercise sessions of 2, 3 and 4 minutes.

8   Repeat steps 4 to 7 for a different subject. Test as many subjects as possible.

9   Using the results you recorded:
    a   calculate an average across all subjects for each exercise condition.
    b   graph the results using a line graph.

WHAT DO YOU THINK?

1   What trend can you see in your results?

2   What trends did other groups see in their results?

3   Using your knowledge of the respiratory system, **explain** your results.

4   **Evaluate** your results – i.e. say whether you think your results are reliable and valid, giving reasons.

5   Suggest two ways to improve your investigation.

CONCLUSION

Write a conclusion that summarises your findings. Make sure it relates to your aim.

INVESTIGATION 2: THE STRUCTURE AND FUNCTION OF THE MOUTH

To explain the action of saliva and salivary amylase in the mouth

PART A: EATING AN APPLE

1 Eat a small piece of apple or other hard fruit.
2 Make careful observations about what each part of your mouth does as you bite, chew and swallow the fruit.
3 Consider what is added to the food as you chew.
4 Write down the functions carried out by the lips, teeth (incisors and canines) and tongue.

1 What is the role of saliva?
2 Why do we have different kinds of teeth?
3 What is the function of the mouth in eating?

Extension

4 Compare the mouth, and in particular the teeth, of a lion, piranha and human. Use a graphic tool to visually represent the similarities and differences. See Figure 2.10.1 for reference.

PART B: EATING A BISCUIT

1 Put a savoury biscuit in your mouth and chew for as long as possible before swallowing.
2 Carefully observe and record any changes in taste as you chew the biscuit.

1 What is the role of saliva?
2 Why did the savoury biscuit develop a sweet taste as you chewed it in your mouth?
3 What substance in your mouth would have caused this change?
4 Can you suggest why dogs and cats do not produce this substance?

▲ FIGURE 2.10.1   The mouths and teeth of (a) a lion, and (b) a piranha

Getty Images/Moment/Pal Teravagimov Photography

Getty Images/Photodisc/Sylvain Cordier

REMEMBERING

1 Match the blood constituent (a–d) to its function (i–iv).

Constituent	Function
a Plasma	i Immune functions
b Red blood cells	ii Fluid medium
c White blood cells	iii Blood clotting
d Platelets	iv Carry oxygen bound to haemoglobin

2 **List** three features of the lungs that make them well suited to the job they do.

3 **State** the word equation for cellular respiration.

UNDERSTANDING

4 **Explain** how and why the digestive systems of the koala and the dingo differ from yours.

5 Why do you think peristalsis occurs in the digestive tract?

6 Account for the difference in thickness of the muscular walls of the atria and ventricles.

7 a **Name** the system shown in the diagram at right.
   b **Name** the organs labelled A, B, C and D.

8 **Explain** what happens to the diaphragm during inspiration and expiration.

9 **Describe** how First Nations Australians shared their knowledge of human and animal systems.

APPLYING

10 If you were shown a blood vessel attached to the heart, **explain** how you would know whether it was an artery or a vein.

11 Some fish have kidneys that always produce a large volume of urine. Other species produce very small quantities of urine. **Predict** whether fish that live in the sea belong to the first or second group.

12 **Describe** the pathway of a urea molecule from the aorta to the outside of the body.

13 **Explain** why blood oozes out of a wound to a vein but spurts out of wound to an artery.

14 **Apply** your understanding of the urinary system to explain why marine turtles drink only sea water and constantly produce very salty tears.

ANALYSING

15 **Explain** how the oesophagus and the small intestine are alike. How do they differ?

16 **Compare** mechanical and chemical digestion.

17 **Design** a summary table for digestion, listing each organ and its function.

18 Where in the body would you expect to find haemoglobin combining with oxygen?

19 How do capillaries differ from other blood vessels in:
   a their function?   b their structure?

EVALUATING

20 Why is it incorrect to say all arteries carry oxygenated blood?

21 Should the process of egesting material from the digestive system be called excretion? **Explain** your answer by referring to the internal and external environments.

22 **Explain** the importance of valves in the heart and veins.

23 Each year, many babies are born with a ventricular septal defect, where there is a hole in the septum dividing the ventricles. **Explain** how this could affect blood being pumped out of the aorta.

CREATING

24 **Predict** the symptoms of a person suffering from an iron deficiency.

25 In the early 1900s, little was understood about the role and importance of vitamins in the human and animal diet. Suggest why a scientist working with rats kept in cages in a laboratory and a doctor working with convicts in a local prison both recognised that some illnesses were caused by specific dietary deficiencies.

BIG SCIENCE CHALLENGE PROJECT

#2

1 Connect what you have learned

In this chapter you have learned about various body systems and their disorders. Create a mind map to show how the information you have learned is connected.

2 Check your thinking

Think about the relationships between body systems and human behaviour. Find out why many health disorders are preventable, but many people do not follow medical advice.

3 Make an action plan

Research a disorder/disease of a body system of your choice. Learn about how the condition is caused and how it is managed.

4 Communicate

Use your knowledge and understanding to create a brochure to warn people about the disorder/disease you have chosen. Your brochure should be suitable for a doctor's waiting room and should explain what steps people could take to prevent this disorder.

Shutterstock.com/Crevis

3 Plant systems

3.1 Requirements of plants (p. 62)

Plants require water, carbon dioxide, oxygen and sunlight.

3.2 Photosynthesis (p. 64)

The Sun is the source of energy for all plants on Earth.

3.3 Systems in plants (p. 66)

Specialised cells, tissues and organs make up plant systems.

3.4 Water transport (p. 68)

Plants transport water from their roots to their leaves.

3.5 Sugar transport (p. 70)

Plants transport sugar from their leaves to other parts of the plant.

3.6 Control of gases (p. 72)

Gas exchange occurs between the air and plants.

3.7 Specialised structures (p. 74)

Plants have specialised structures allowing them to live in many different environments.

3.8 FIRST NATIONS SCIENCE CONTEXTS: First Nations Australians' knowledge and use of plants (p. 77)

First Nations Australians have always harvested plants and applied their traditional knowledge to support sustainability of plant resources.

3.9 SCIENCE AS A HUMAN ENDEAVOUR: Controlling land clearing with native vegetation clearance controls (p. 79)

Clearance controls can reduce habitat destruction and minimise the risk of reduced biodiversity.

3.10 SCIENCE INVESTIGATIONS: Developing questions, predictions and hypotheses (p. 80)

1 Investigating factors affecting stomata density
2 Describing water transport in a plant

▲ FIGURE 3.0.1 Different types of plants are suited to different environments.

Is this a typical plant? What kind of environment does it live in? How do you know?

Plants are a very diverse group of organisms that inhabit a wide variety of environments, from deserts to rainforests, to grassy plains and marine environments. What features might help a plant get enough water, sunlight, carbon dioxide and oxygen, no matter where they are living?

▶ Can you find out more about what has enabled the spread and success of plants in such a wide variety of climates and conditions?

#3 SCIENCE CHALLENGE ACCEPTED!

At the end of this chapter, you can complete Big Science Challenge Project #3. You can use the information you learn in this chapter to complete the project.

Assessments
- Prior knowledge quiz
- Chapter review questions
- End-of-chapter test
- Portfolio assessment task: Data test

Videos
- Science skills in a minute: Questions, predictions and hypotheses (3.10)
- Video activities: Respiration in plants (3.1); Photosynthesis (3.2); Plant transport (3.5); Land clearing (3.9)

Science skills resources
- Science skills in practice: Developing questions, predictions and hypotheses (3.10)
- Extra science investigations: Plant growth (3.1); Photosynthesis (3.2); Seed germination (3.7)

Interactive resources
- Label: Plant systems (3.3); Leaf cross-section (3.6)
- Drag and drop: Plant requirements (3.1); Water transport in plants (3.4)
- Crossword: Plant systems (3.4)

✸ Nelson MindTap

To access these resources and many more, visit:
cengage.com.au/nelsonmindtap

Video activity
Respiration in plants

Interactive resource
Drag and drop: Plant requirements

Extra science investigation
Plant growth

Plants require water, carbon dioxide, oxygen and sunlight

The chemical reactions that make plants grow occur in the watery cytoplasm of the plant's cells. This means that sufficient water is an important requirement of plant cells. Water is also used by plants to transport materials from the roots to the leaves, from the leaves to the roots and between all other parts of the plant.

Two vital processes that take place in plant cells are photosynthesis and cellular respiration. These two processes require water, carbon dioxide, oxygen and sunlight (Figure 3.1.1). Cellular respiration in plants is the same as cellular respiration in animals, which you learned about in Chapter 2. Photosynthesis will be discussed further in Module 3.2.

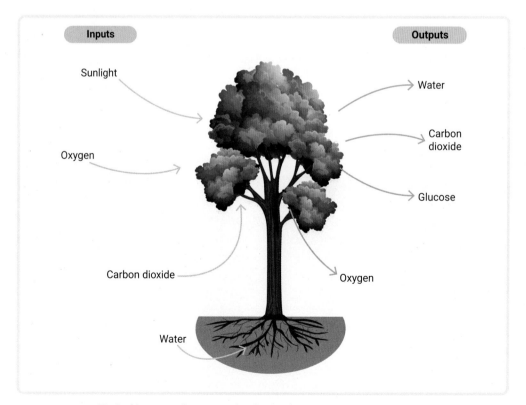

▲ FIGURE 3.1.1   Typical inputs and outputs of a plant

Minerals required by plants

Carbon, oxygen and hydrogen are elements that are present in carbon dioxide and water. However, plants also require phosphorus, potassium, nitrogen, sulfur, calcium, iron and magnesium. These are supplied by minerals in the soil. A shortage of one or more of these minerals will affect the growth and appearance of plants. For example, a lack of nitrogen in the soil will cause reduced plant growth. If there is not enough nitrogen, magnesium, zinc or iron, the plant's leaves may turn yellow because these minerals help to make the green pigment chlorophyll. You will learn more about this in Module 3.2.

nodules
swellings on the roots of plants

Even though Earth's atmosphere is 78 per cent nitrogen, plants cannot absorb this element from the air. Instead, they get their nitrogen from the soil. In the atmosphere, nitrogen is a gas. In the soil, it is in the form of a salt called a nitrate. Some bacteria living in the soil are able to convert atmospheric gaseous nitrogen into this salt form. These bacteria can also live in the roots of legumes, such as beans and peas. Here, they cause the root to swell and form **nodules** (Figure 3.1.2).

▲ **FIGURE 3.1.2** Nodules on roots of a soybean plant

Some plants have evolved a different way to ensure they obtain enough nitrogen: they have become carnivorous. The Australian tropical pitcher plant has a leaf shaped like a pouring jug (also known as a pitcher). The colour and sweet smell of the leaf attracts insects. When insects land on the rim of the slippery pitcher, they slide in and cannot escape. Inside the pitcher plant, the insect gradually breaks down and the plant absorbs its nutrients (Figure 3.1.3). Carnivorous plants often live in boggy soils that do not have much nitrogen.

Comparing plant and animal requirements

Animals require water, food and oxygen to survive. Animals rely on plants or other animals for their food, which supplies them with minerals, vitamins and other important substances for survival and good health.

Plants similarly need water and oxygen for respiration, but they also need carbon dioxide and sunlight to power photosynthesis so that they can make their own food. Plants require a range of minerals to function, survive and thrive.

▲ **FIGURE 3.1.3**
A pitcher plant

3.1 LEARNING CHECK

1 **List** the substances that plants need to live. Describe the processes that require these substances.
2 **Explain** why pitcher plants need to catch and digest insects.
3 **Describe** how root nodules benefit a pea plant.
4 **Compare** the requirements for plants and animals to live.

BY THE END OF THIS MODULE, YOU WILL BE ABLE TO:

✓ describe the process of photosynthesis

✓ compare and contrast photosynthesis and cellular respiration.

Video activity
Photosynthesis

Extra science investigation
Photosynthesis

> **GET THINKING**
>
> Look at the glossary terms in the margins in this module. Are you familiar with some of the words? Are there any words that you would like to know more about?

Photosynthesis

Green plants use light energy from the Sun to make their own food in a process called photosynthesis. In this process, carbon dioxide and water combine to form glucose (a simple sugar) and oxygen. The plant uses the glucose for cellular respiration and other purposes to ensure its survival. It expels the oxygen. The chemical reactions that make up photosynthesis are summarised in the following word equation.

$$\text{carbon dioxide} + \text{water} \xrightarrow[\text{chlorophyll}]{\text{sunlight}} \text{glucose} + \text{oxygen}$$

Some plants have lost the ability to photosynthesise. Most of these are parasitic and obtain their nutrients from underground fungi.

The site of photosynthesis

chloroplast
a green organelle in plant cells that contains chlorophyll and carries out photosynthesis

chlorophyll
the green pigment in chloroplasts that absorbs light for photosynthesis

The cells in plants that carry out photosynthesis contain around 30 to 40 organelles called **chloroplasts**. These organelles are oval-shaped and contain the green pigment **chlorophyll**. In photosynthesis, chlorophyll absorbs light energy, converting it to chemical energy that is stored in glucose. All of the green parts of a plant, usually the leaves and sometimes the stems, contain chloroplasts and carry out photosynthesis (Figure 3.2.1).

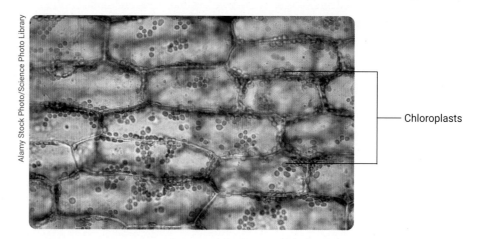

Alamy Stock Photo/Science Photo Library

Chloroplasts

▲ **FIGURE 3.2.1** Plant leaf cells containing chloroplasts. The green colour is due to chlorophyll in the chloroplasts.

Comparing photosynthesis and respiration

Take a look at the equations for respiration and photosynthesis. How do they compare?

Respiration: glucose + oxygen $\longrightarrow$ water + carbon dioxide + energy

Photosynthesis: carbon dioxide + water $\xrightarrow[\text{chlorophyll}]{\text{sunlight}}$ glucose + oxygen

In terms of the substances used and produced, they are the opposite of each other. In terms of energy, it is *produced* in respiration while it is *required to drive* photosynthesis. While the goal of photosynthesis is to store energy by building glucose from carbon dioxide and water, respiration is the reverse of this. Glucose is converted back into carbon dioxide and water, and energy is released.

Phytoplankton as a carbon sink

Phytoplankton are microscopic, unicellular organisms that photosynthesise in the same way as plants (Figure 3.2.2). Phytoplankton produce at least half of the oxygen on Earth. They live in the upper sunlit layer of almost all oceans and bodies of fresh water.

Since phytoplankton photosynthesise, they absorb huge quantities of carbon dioxide from the atmosphere. They are also a source of food for a wide range of organisms, including shrimp, snails and jellyfish. These organisms, in turn, are eaten by fish, dolphins and whales. Phytoplankton are therefore the foundation of all aquatic food webs. If phytoplankton die before they are eaten, their bodies sink to the sea floor. They remain there as a store of carbon, often referred to as a **carbon sink**. The oceans store 50 times more carbon – as carbonates in coral and sea shells and dissolved carbon dioxide – than the atmosphere.

phytoplankton
microscopic, photosynthetic organisms that live in the sea or in fresh water

carbon sink
a natural or artificial storage place for carbon, for an indefinite period

▲ **FIGURE 3.2.2**   A range of microscopic phytoplankton

3.2 LEARNING CHECK

1 **Define** 'carbon sink' and **describe** its role.
2 **Describe** the relationship between chlorophyll and a chloroplast.
3 **Describe** the importance of phytoplankton in reducing the high levels of carbon dioxide in the atmosphere.
4 **Explain** whether photosynthesis is the opposite of respiration.
5 **Explain** why photosynthesis is vital to a plant's survival.

3.3 Systems in plants

BY THE END OF THIS MODULE, YOU WILL BE ABLE TO:

✓ describe the structure and function of plant systems.

Interactive resource
Label: Plant systems

GET THINKING

When you eat a salad, it could contain many different plant tissues and organs, including seeds, fruit, leaves, stems, roots and even flowers. How do plant tissues, organs and systems work together in a plant?

shoot system
the leaves, stems and flowers of a plant; usually above ground

Cells, tissues, organs and systems in plants

As we saw in Chapter 1, in a multicellular organism such as a plant, many cells combine to produce one individual. Because of its size, a plant cannot rely on diffusion to transport its requirements to each of its cells. In plants, as in animals, we see a division of labour. When cells differentiate and become specialised to perform a particular function, they form tissue. Tissues perform important functions that support the life of the plant. These include obtaining energy, distributing materials, removing waste and exchanging gases. Sometimes different tissues are organised to form an organ, such as a leaf.

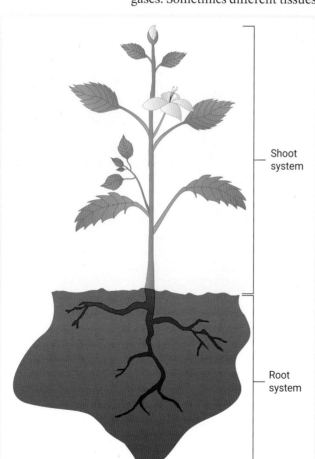

In Chapter 2 we looked at some of the important organ systems of animals. Plants do not have all these systems. There are only two organ systems in plants (Figure 3.3.1). These are the:

- shoot system – the stems, leaves and flowers
- root system – all parts of the roots.

Shoot system

The **shoot system** is generally the part of the plant that is above ground. It consists of the leaves, stems and flowers. In trees, the stem is the trunk and all the branches, including the smallest twigs. The stems of some plants are green and can photosynthesise. However, the leaves are the major organs of photosynthesis. In many plants, the leaf consists of a flattened blade and a stalk that attaches the leaf to the stem. The stem supports and spaces out the leaves, to ensure they have good access to sunlight and carbon dioxide.

▲ **FIGURE 3.3.1** The two organ systems of a plant are the shoot system and the root system.

Root system

The **root system**, consisting of the roots and root hairs, is usually below ground. The root system anchors the plant in the soil, absorbs and transports water and minerals, and stores food. Roots have thousands of hairs just behind their tip. These hairs increase the surface area of the root and allow it to quickly absorb water and nutrients (Figure 3.3.2).

The shape of plant roots varies. Plants with **tap roots** have a large tapering main root with slender, side branches. Vegetables such as carrots, turnips and sweet potato have large tap roots, which are used to store food.

Some plants, such as grasses, palms and sugarcane, have **fibrous roots**. These fibrous roots consist of a mat of smaller thread-like roots of similar size. Although fibrous roots do not grow very deep, they hold the soil strongly in place (Figure 3.3.3).

▲ FIGURE 3.3.2 Root hair cells greatly increase the surface area for absorption of water.

root system
the water- and nutrient-absorbing part of a plant; usually below ground

tap root
the large tapering main root of some plants

fibrous roots
many small roots of similar size that grow from the bottom of the stem of some plants

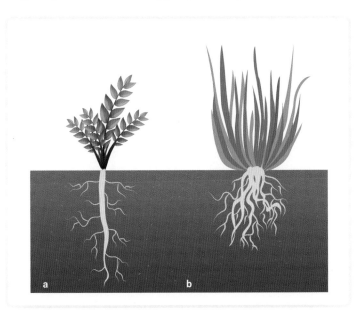

▲ FIGURE 3.3.3 (a) A typical taproot system; (b) a typical fibrous root system

 LEARNING CHECK

1 **Name** and **describe** the two systems that make up a plant.
2 **State** the functions of the two systems in plants.
3 **Compare** tap roots and fibrous roots.
4 **Explain** why the root hair cells in Figure 3.3.2 are called tissue.
5 Which system in an animal is like the root system in a plant? **Explain** your reasoning.

BY THE END OF THIS MODULE, YOU WILL BE ABLE TO:

✓ describe the transport of water in plants.

vascular bundle
a combined strand of xylem and phloem tissue in plants

vein
a vascular bundle of xylem and phloem tissue in a leaf

xylem
plant tissue that transports water and minerals from the roots to the rest of the plant

lignin
a material that stiffens and strengthens plant cell walls

GET THINKING

Your heart pumps blood to all your body cells. How is water moved from plant roots to all parts of the shoot without a pump?

Water conducting tissues

For photosynthesis to occur, the leaves of a plant need maximum exposure to available sunlight, so the best position for a leaf is high above the ground, at the uppermost tip of the stem. Water must travel from the plant's roots to where it is needed in the leaves, and this can be a long way. For example, the tallest gum tree, the Centurion tree in Tasmania, is 100 metres tall.

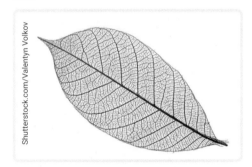

▲ **FIGURE 3.4.1** The branched network of veins in a leaf

Plants transport substances in **vascular bundles**. In the leaves, the vascular bundles are called **veins**. The veins may form a highly branched network (Figure 3.4.1), or they may be parallel, as seen in grass species.

Vascular bundles are made of two types of tissue, xylem and phloem. Phloem transports sugar around the plant – this will be discussed in Module 3.5. **Xylem** carries water from the roots of plants to their leaves. It is composed of groups of long, thin tubes like pipes, made from dead xylem cells with thick walls (Figure 3.4.2).

Lignin is a complex material that makes plant cell walls more rigid. Xylem cells contain lignin, which makes them strong and stiff. Wood is composed almost entirely of xylem, strengthened with lignin, and provides the main support for large plants such as trees.

The pathway of water through the plant

Root hairs provide a large surface area that allows plants to absorb water. The root hair cells grow between the soil particles so that they are in close contact with water in the soil. Root hair cells also need to be near air pockets in the soil (Figure 3.4.3). This gives them access to oxygen, one of the substances necessary for cellular respiration, which produces energy. Some of this energy is used to pump minerals into the roots from the soil.

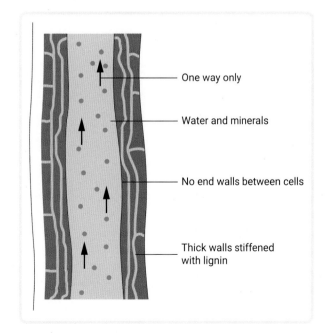

One way only

Water and minerals

No end walls between cells

Thick walls stiffened with lignin

▲ **FIGURE 3.4.2** Xylem transports water from the roots to the leaves.

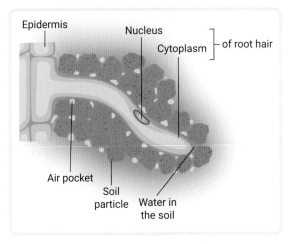

▲ FIGURE 3.4.3   A root hair cell in detail

Shutterstock.com/Carolina K. Smith MD

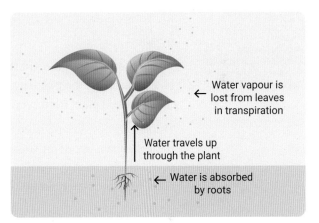

▲ FIGURE 3.4.4   Internal stem structure, showing the vascular bundles arranged in a circle

Once inside the root hair, the water moves into the xylem tissue located in the vascular bundle in the root. The water then travels to the rest of the plant.

In the stem, the vascular bundles are arranged in a circle, with the xylem tissue on the inside (Figure 3.4.4). Every leaf cell is close enough to the vascular bundles of the plant to ensure it has enough water. A continuous column of water runs through the xylem tissue up the stem of the plant to the leaves (Figure 3.4.5). This is quite incredible, because there is no machinery in a plant to pump the water upwards. Water can do this because of the way its molecules stick together. Water molecules are attracted to each other, so when one molecule moves up, it pulls the other molecules with it.

▲ FIGURE 3.4.5   The movement of water in the xylem from the roots to the leaves

In Chapter 1 you learned how the guard cells on a leaf can prevent water loss by evaporation, by closing the pore between them. In Module 3.6 you will learn more about guard cells. When water molecules move from the leaf into the air, other water molecules move up the xylem to replace those that were lost. This process is called **transpiration**, and it is how plants can draw water out of the soil and move it all the way up to the highest leaves.

transpiration
the evaporation of water from the leaves of a plant

3.4 LEARNING CHECK

1 **Define**:
   a vascular bundle
   b vein.
2 **Describe** the structure of xylem tissue.
3 Why do root cells need oxygen?
4 What is transpiration and what is its purpose?
5 **Explain** why the structure of a root hair cell is suited to its function.

Interactive resources
Drag and drop: Water transport in plants
Crossword: Plant systems

BY THE END OF THIS MODULE, YOU WILL BE ABLE TO:

✓ describe how plants transport sugar from the leaves to other parts of the plant.

Video activity
Plant transport

> **GET THINKING**
>
> In the previous module, you learned how water moves from the roots of plants to the leaves. How do you think sugar is transported around the plant?

The function of phloem

Not all of the glucose produced by photosynthesis is used by leaf cells for cellular respiration. Some of the glucose is converted to **sucrose**, which is what we know as sugar. **Phloem** is the tissue responsible for transporting sucrose to other parts of the plant. As we saw in Module 3.4, phloem, together with xylem, occurs in the vascular bundles.

Phloem sap is very different from the watery sap in the xylem. High concentrations of sucrose make the phloem sap thick and syrupy (sugary!). The sap may also contain minerals and plant hormones.

Phloem carries sucrose from the leaves to any part of the plant where it is needed, either for immediate use or for storage. Roots, shoot tips, stems, flowers and fruit all need sugar in order to grow. Any extra sucrose is carried to the root system of the plant, where it is converted into starch for long-term storage.

sucrose
the form of sugar transported in the phloem of a plant

phloem
plant tissue that transports sucrose from the leaves to the rest of the plant

The structure of phloem

Phloem cells, called **sieve tube cells**, are tubular, like xylem cells. Their end walls are perforated by holes to form **sieve plates** (so called because they look like a sieve). Sugar moves from one sieve tube cell to the next through the sieve plates. Unlike xylem tissue, phloem tissue is made of living cells (Figure 3.5.1).

sieve tube cell
a food-conducting cell in a plant that forms phloem

sieve plate
holes at each end of a sieve tube cell that allow the passage of sucrose

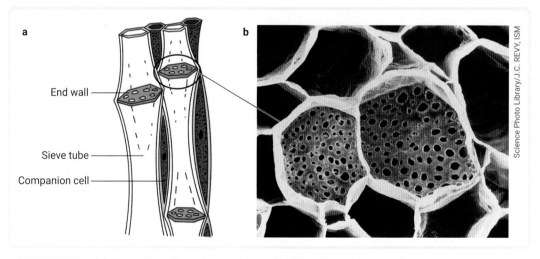

▲ **FIGURE 3.5.1** **(a)** Sieve tube cells and companion cells; **(b)** a sieve plate, seen from above

Transport in the phloem tissue is very specific. One sieve tube in a vascular bundle may carry sap in one direction, while another sieve tube in the same vascular bundle may carry sap in the opposite direction. Sieve tube cells are closely associated with adjacent **companion cells**. Without its companion cell, the sieve tube cell would die.

Ring barking

Ring barking is the removal of bark, which contains the phloem, from around the entire circumference of the main trunk of a tree. This may happen through accidental damage, grazing by animals or sometimes insect attack. Sometimes it can also be done intentionally to kill the tree.

Because the phloem is on the outside of the vascular tissue (Figures 3.4.4 [page 69]and 3.5.2), ring barking prevents movement of sugar through the phloem. Since the xylem is undamaged, water and nutrient uptake and transport continues unchanged.
At first, there is enough stored starch in the roots to maintain cellular respiration and root growth. However, over time these reserves are used up and the tree dies.

companion cell
in plants, a cell adjacent to a sieve tube cell, which makes substances unable to be made by the sieve tube cell

ring barking
a way of killing a tree by removing the bark containing phloem from around the trunk, but leaving the xylem intact

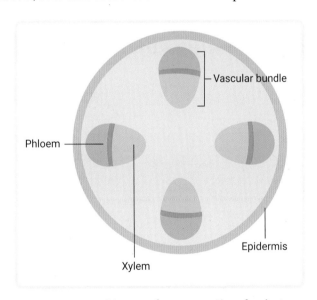

▲ **FIGURE 3.5.2**   Diagram of a cross-section of a plant stem showing phloem on the outside

3.5

3.5 LEARNING CHECK

1 **Describe** the difference between xylem sap and phloem sap.
2 **Name** the cells that are responsible for keeping sieve tube cells alive.
3 **State** what phloem tissue is used for.
4 **Explain** why sucrose is transported to the fruit of a plant.
5 **Explain** why a ring-barked tree dies over time, rather than immediately.
6 Give two reasons why sugar, produced in the leaves by photosynthesis, is transported to plant root cells.

3.6 Control of gases

BY THE END OF THIS MODULE, YOU WILL BE ABLE TO:

- ✓ describe how gas is exchanged between plants and the air
- ✓ explain how plants control water loss from their leaves.

GET THINKING

Look at the glossary terms in the margins in this module. Are you familiar with some of the words? Are there any words that you would like to know more about?

Tissues of the leaf

epidermis
the cellular surface layer of a plant

cuticle
the waxy protective layer on the surface of a plant

The leaf is the site of photosynthesis. Although leaves vary in size and shape, they are generally flat and thin. The **epidermis** is a single layer of cells on the upper and lower surface of the leaf. The cells in the epidermis are tightly packed to help prevent water loss from the leaf and protect the plant from attack by bacteria and fungi. The epidermis is covered by a waxy layer, called the **cuticle**, which also reduces water loss (Figure 3.6.1).

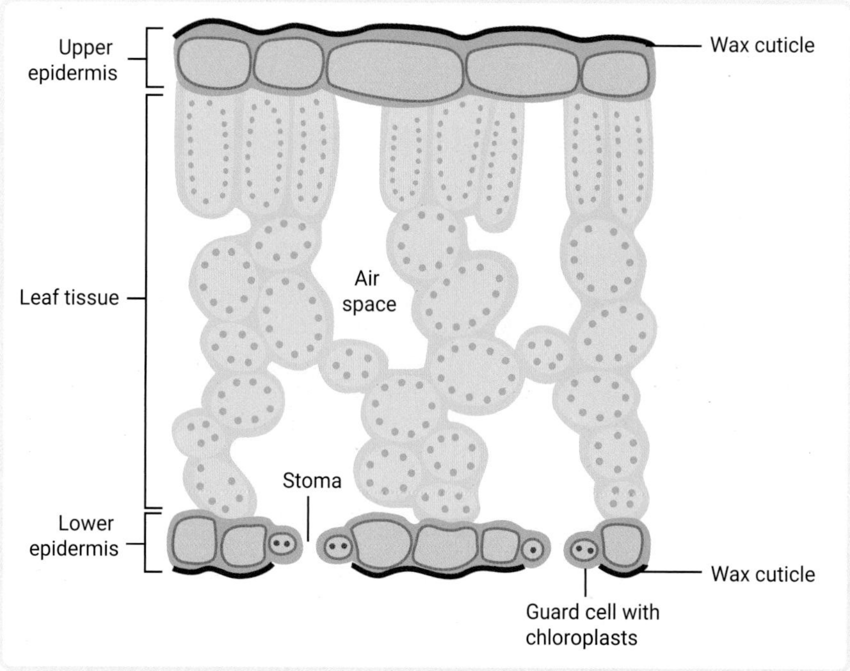

▲ **FIGURE 3.6.1** Cross-section of a typical leaf. Chloroplasts are shown as green dots inside most of the leaf cells.

The main leaf tissue is between the upper and lower epidermis. At the top of the leaf the cells are brick-shaped and fairly closely packed. Below them are loosely arranged, irregular cells, with many air spaces between them, which contain many chloroplasts.

Gas exchange in the leaf

When the leaf is photosynthesising, it takes in carbon dioxide and produces oxygen. These gases enter and exit the leaf through openings in the epidermis called stomata (singular: stoma). There are usually more stomata on the lower surface of the leaf – where they are protected from the heat of the Sun – than on the top surface.

Interactive resources
Label: Leaf cross-section
Crossword: Plant systems

Extra science investigation
Seed germination

The air spaces within the leaf are large near the stomata. This arrangement allows the carbon dioxide that enters the leaf through the stomata to move freely to all of the leaf cells. It also allows oxygen to easily move from the cells where it is produced in photosynthesis to the outside through the stomata.

Guard cells are pairs of crescent-shaped cells that surround each stoma. When the guard cells absorb water, they swell. This has the effect of opening the stoma. When the guard cells lose water, they flop, causing the stoma to close (Figure 3.6.2). Although the opening and closing of stomata can be affected by several factors, stomata are usually open during the day and closed at night (Figure 3.6.3). Guard cells are the only cells in the epidermal layer to contain chloroplasts.

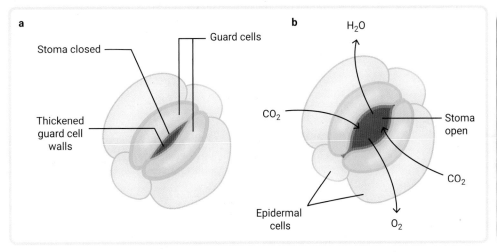

▲ FIGURE 3.6.2   (a) Stoma closed; (b) stoma open: carbon dioxide can enter the leaf; water vapour (water in gas form) and oxygen escape

Movement of water through the leaf

The vascular bundles of the stem link up with those in the leaf. In Module 3.4, you saw that xylem in the vascular bundles carries water to the leaf, ensuring a ready supply to keep the inside of the leaf moist. During hot weather, a plant may lose water through transpiration from its leaves faster than the xylem can supply it. High water loss causes a plant to **wilt**, where it becomes floppy and droopy. This is where the ability of plants to open and close their stomata is important. As a plant experiences high water loss and wilting, the guard cells also lose water, causing the stomata to close. This reduces further water loss. With enough water, a wilted plant may recover.

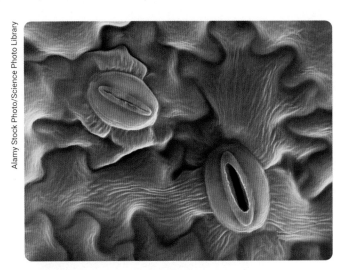

Alamy Stock Photo/Science Photo Library

▲ FIGURE 3.6.3   A surface view of a leaf showing stomata: one open and one closed

wilt
to become limp and to droop through loss of water

Quiz
Specialised plant structures

GET THINKING

Think of some examples of plants that you know can live in extreme environments. How do they live there successfully? How are they different from other plants?

Living in water

A problem faced by underwater aquatic plants is that there is less carbon dioxide and oxygen in water than in air. Therefore, an important adaptation of aquatic plants is that they are not covered by a waxy cuticle. This means gas exchange occurs directly between their photosynthetic cells and the water.

estuary
an area where a freshwater river meets the ocean

Mangroves live in waterlogged soils of coastal wetland **estuaries**. They are the only trees that thrive in these areas because of their ability to survive in both salt and fresh water. When they absorb water, their roots can block some of the salt and they excrete highly saline water from specialised salt glands in the epidermis of their leaves.

Mangroves solve the problem of a lack of oxygen by having aerial roots, called peg roots, that link with underground roots. Numerous pores on the surface of the peg roots allow gaseous exchange even when the underground roots are under water due to the tide (Figure 3.7.1).

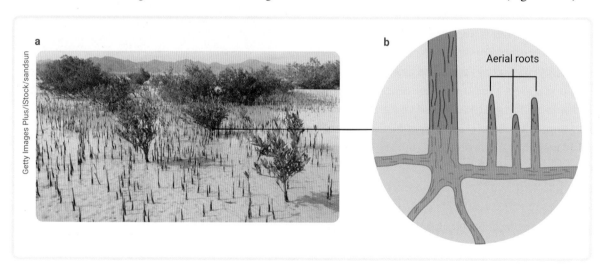

▲ **FIGURE 3.7.1** **(a)** Mangroves and their aerial roots; **(b)** a closer look at aerial roots

Rainforest plants

canopy
the 'top' of a forest, made up of overlapping leaves and branches of tall trees

Rainforests have a thick, almost continuous, tree **canopy**. It is made up of the overlapping leaves and branches of all the tall trees. The conditions of low light levels and still, moist air beneath the canopy are very different from those outside the rainforest. Plants growing on the floor of the forest usually have very large, flat, dark green leaves (Figure 3.7.2). This type of adaptation increases the surface area available for light absorption in the dim conditions.

The top 15–20 centimetres of soil on the rainforest floor is decaying leaves, wood and other organic matter. Rainforest plants generally have shallow roots to tap into this rich pool of nutrients.

Plants in hot, dry environments

For plants where water is readily available, such as in a rainforest, the opening and closing of stomata is enough to maintain water balance. But many plants live in hot, dry places where they must reduce their water loss in other ways. High daytime temperatures, combined with low humidity (water in the air), increase the rate of transpiration and cause water to evaporate through the leaf's epidermis and cuticle. Often, there is little water in the soil to replace this loss. Some plants that live in areas of low rainfall and extreme temperatures have a variety of adaptations to help them survive.

Leaf size and type

In contrast to the large-leaved plants in rainforests, plants in hot, dry climates often have small leaves. Small leaves mean less surface area and fewer stomata to allow water to escape during transpiration. The leaves also have a very thick, waxy cuticle and multiple layers of epidermal cells to reduce water loss from the leaf surface (Figure 3.7.3). Some plants have shiny leaves that reflect heat and hard, leathery leaves that reduce damage caused by wilting.

Stomata

A number of plant species open their stomata at sunrise and close them in the middle of the day to reduce water loss. They reopen the stomata in the cool of evening. In other species, the stomata may occur in pits on the underside of the leaf; this is known as 'sunken' stomata. This arrangement reduces water loss, but still allows gas exchange for photosynthesis.

Roots

Many plants in dry areas have long tap roots to reach deep into the soil to access water. Some also have an extensive surface root system to absorb any water present. First Nations Australians have long known how to extract water from these types of roots.

▲ FIGURE 3.7.2   Plants growing below the canopy in a rainforest often have large, flat leaves.

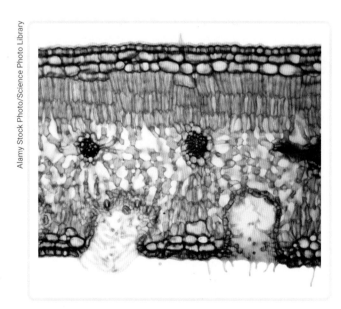

▲ FIGURE 3.7.3   A microscopic cross-section of a plant leaf showing adaptations to a hot dry environment: a thick waxy cuticle, multiple layers of epidermal cells and 'sunken' stomata

Shortened life cycles

There are plants that cope with drought and high temperatures by not existing during either! These drought-avoiding plants are called **ephemerals**, which is a word meaning 'lasting for a short time' (Figure 3.7.4). Ephemerals complete their life cycles quickly and spread large quantities of seeds widely. These seeds can resist heat and drought for long periods, remaining in the soil until rain triggers germination.

ephemeral
describes a plant completing its life cycle quickly and spreading large quantities of seeds

▲ **FIGURE 3.7.4** An ephemeral in flower in the desert

Adaptations to fire

Although some plants are killed by bushfires, one benefit of fires is that they open up spaces and create a nutrient-rich seedbed. Many species benefit from fire, including bottlebrushes, hakeas, some acacias and eucalypts. Thin-barked trees rely on their seeds being protected in hard wood capsules, called gum nuts, which are released after the fire has moved through.

Eucalypts have several adaptations that help them recover from fire (Figure 3.7.5). Thick bark protects some species from the extreme heat. High-intensity fires can stimulate buds beneath bark on large branches to begin growing. They appear as bright green, bushy foliage. Some eucalypts have a large swelling at their base, known as a **lignotuber**. The lignotuber holds buds and nutrients, allowing the tree to vigorously re-sprout after fire.

lignotuber
a partly underground swelling of the trunk of a plant, with many buds that sprout after fire

 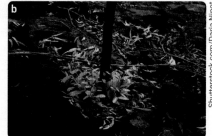 

▲ **FIGURE 3.7.5** **(a)** The mountain ash tree. Bushfires readily kill this eucalypt, but before it dies it releases millions of seeds. **(b)** Leaves sprout from lignotubers at the base of a eucalyptus tree after fire. **(c)** After damage by fire, buds just below the bark are stimulated to produce shoots.

3.7 LEARNING CHECK

1 **State** three ways in which plants in hot, dry areas reduce water loss from their leaves.
2 **Describe** two adaptations of mangroves to estuaries.
3 **Explain** the benefit of:
   a a lignotuber.
   b stomata in pits.
   c large, broad leaves in a rainforest.
   d deep tap roots.
4 **Explain** how ephemerals survive in hot, dry environments.
5 **Describe** two ways of plants have adapted to fire.

9780170463027

3.8 First Nations Australians' knowledge and use of plants

IN THIS MODULE, YOU WILL:

✓ recognise that First Nations Australians have successfully and sustainably harvested plant materials for thousands of years

✓ examine examples of how First Nations Australians applied their traditional knowledge to support sustainability of plant resources.

Harvesting plant materials sustainably

First Nations Australians accumulated a wealth of information and experience concerning the biology, ecology and physical properties of plants over many thousands of years. Although plants form a crucial part of the traditional diet, they were also collected and harvested for many other purposes. As well as being used as a source of medicines and water, plants are important in the construction of tools, weapons, ornaments, musical instruments and toys, and even as an aid to catching animals.

Because plants were essential, First Nations Australians understood that it was important to harvest plant resources sustainably, so that future generations would still have them.

Harvesting bark

The outer bark of specific types of trees has been carefully harvested by First Nations Australians for millennia to be used in the construction of canoes, shields, tools, implements and as a canvas for painting or to expose heartwood for carving. To ensure the ongoing survival of the tree, great care was taken not to cause too much damage to the plant's system when the bark was removed. Across southern and eastern Australia, you can still see the presence of living, scarred trees. This shows First Nations Australians' in-depth knowledge of plant biology.

The Alyawarre Peoples of the Central Desert Region carefully cut pieces of outer bark from opposite sides of the ghost gum (*Corymbia aparrerinja*) to make trays that are used to process resin or gum. In the Hunter Valley region

▲ **FIGURE 3.8.1** A canoe scar-tree in Yugambeh Country, south east Queensland. Careful removal of the outer bark to construct a canoe ensures the tree's survival.

◀ **FIGURE 3.8.2** A shield tree that was harvested to make a Kaurna shield, South Australia.

◀ **FIGURE 3.8.3** The bark from trees such as the paperbark can be used to make blankets.

of New South Wales, there is a very old tree that bears scars believed to be more than 100 years old, and is consistent with the removal of bark to manufacture canoes. The Dyirbal Peoples of the north Queensland rainforest region have long collected bark up to 12 metres above the ground from the banana fig (*Ficus pleurocarpa*) to prevent damage to the root system and ensure survival of the tree. The bark was used to manufacture blankets.

☆ ACTIVITY 1

1 Use what you have learned about plant systems to **explain**:
   a which plant system(s) is being affected by the removal of bark.
   b why the careful removal of bark does not kill the tree.
   c why a scar forms.

Sustainable cultivation

First Nations Australians have long used controlled fires to promote the growth, propagation and germination of certain plants and seeds. This agricultural technique is known as fire-stick farming.

Historically, cycads have been an important source of carbohydrates for many First Nations Australians cultural groups. Carefully controlled fires are used to promote the distribution and growth of cycads. In certain areas, discrete groves of cycads are grown and fire is used to improve the quality and yield of crops as well as trigger fruit production when required.

Like the cycads, many other Australian native plants have evolved to become fire tolerant. Some species of acacia and banksia require the heat and/or smoke of a fire to start the germination of seeds. Other species such as eucalypts have developed thick bark that covers the lower sections of their trunks.

First Nations Australians' knowledge of the adaptations of these plants allowed them to use fire-stick farming to promote growth, distribution and regeneration of these plants.

▲ FIGURE 3.8.4 A grove of cycads, Belyuen, NT

☆ ACTIVITY 2

1 **Describe** how fire-stick farming is similar to traditional Western farming techniques.
2 a **Explain** what adaptations plants need to thrive by fire-stick farming.
   b **Describe** what would happen to plants that don't have these adaptations.
3 **Suggest** why controlled low-temperature fires are used by First Nations Australians when fire-stick farming.

9780170463027

3.9 Controlling land clearing with native vegetation clearance controls

BY THE END OF THIS MODULE, YOU WILL BE ABLE TO:

✓ describe the role of native vegetation clearance control laws in maintaining biodiversity.

Habitat destruction is pushing some species towards extinction. Since colonisation, large areas of Australia have been cleared for agriculture, grazing and urban development. This has destroyed the habitats of native plants and animals and reduced **biodiversity**. When people clear the land, they often leave only small, separated areas of natural vegetation. This causes problems because many species rely on being able to move between habitats. Land clearing also threatens native birds, many of which need nesting hollows that occur only in very old trees.

Deep-rooted trees keep the water table levels low, preventing salt in the soil from rising to the surface in water. When such trees are removed and agriculture takes over, the soil becomes more salty. Erosion is also more likely to occur.

In 1983, South Australia introduced state-wide native vegetation clearance control laws. Despite some early opposition from farmers and graziers, these laws are now considered very important by the community. Most other states now have such laws.

There are many benefits to limiting vegetation clearance. Photosynthesising plants remove the greenhouse gas carbon dioxide from the air. Linked areas of vegetation, known as wildlife corridors, reduce habitat fragmentation, allowing species to access larger areas (Figure 3.9.1). Protecting native vegetation helps protect endangered animals and rare plants such as orchids.

However, there is more work to be done. The biggest threat to koalas in Queensland, for example, is habitat destruction (Figure 3.9.2). With rapid human population growth and an increased need for housing, koala habitats are becoming smaller and more fragmented. Satellite images help monitor these areas and detect illegal land clearing.

biodiversity
the variety of living species on Earth, including plants, animals, bacteria and fungi

▲ **FIGURE 3.9.1** A wildlife corridor linking two areas of natural vegetation

▲ **FIGURE 3.9.2** Koalas are being affected by vegetation clearance.

3.9 LEARNING CHECK

1 **Explain** the role of satellite images in protecting biodiversity.
2 How does vegetation clearance affect global warming?
3 **Describe** the threats facing koalas in Queensland.

Video activity

Land clearing

Developing questions, predictions and hypotheses

SCIENCE SKILLS IN FOCUS

IN THIS MODULE, YOU WILL FOCUS ON LEARNING AND IMPROVING THESE SKILLS:

▶ develop questions, predictions and hypotheses

▶ describe factors affecting stomata density

▶ describe transport in the plant stem.

CONSIDER THE FOLLOWING STEPS WHEN PLANNING YOUR INVESTIGATION.

▶ **Pose a question**

• The research question describes the purpose of your investigation. What question should you ask to achieve the purpose of your investigation? A question must be able to be investigated.

• An example is: Does temperature affect the rate of photosynthesis?

▶ **Make a prediction**

• A prediction is a statement about what you think will happen regarding your question or your investigation. This prediction should be based on your scientific knowledge and not just a guess.

• An example is: High temperatures will increase the rate of photosynthesis.

▶ **Develop a hypothesis**

• The hypothesis is an educated guess, based on what you already know or have observed. Your hypothesis should be simple and specific. It is a statement that predicts the effect of changing the independent variable (such as temperature) on the dependent variable (the rate of photosynthesis). To make it a fair test, you need to keep all other variables the same. You cannot prove a hypothesis. Your results might support or disprove your hypothesis.

• An example is: The higher the temperature, the greater the rate of photosynthesis.

Video
Science skills in a minute: Questions, predictions and hypotheses

Science skills resource
Science skills in practice: Developing questions, predictions and hypotheses

INVESTIGATION 1: INVESTIGATING FACTORS AFFECTING STOMATA DENSITY

AIM

To compare stomata density on different plants or different parts of a plant

PART A: LEARNING HOW TO COUNT STOMATA

YOU NEED

☑ plant leaf
☑ fine forceps
☑ clear nail polish
☑ microscope
☑ microscope slide

WHAT TO DO

1 Paint an area of approximately 5 mm × 10 mm on the underside of the leaf with the nail polish.

2 Once it is fully dried and set, peel it off using fine forceps and place it on the microscope slide.

3 Examine it under the microscope using 100 × magnification.

4 Count the number of stomata in the field of view. Record your count.

5 Move the slide to a different area of the sample and count the number of stomata again. Record your count.

6 Repeat Step 5 once more.

WHAT DO YOU THINK?

1 Compare the number of stomata in your three different areas.

2 Comment on the consistency of your results.

 Warning

Carry the microscope with two hands and always use it on a level surface.

The glass slides are sharp. Take care when using them. Report any breakages to your teacher immediately.

PART B: DESIGN YOUR OWN INVESTIGATION

Investigate the density of stomata on leaves using one of the following:

- different parts of the same plant
- different plants of the same species in different environments
- species with adaptations to limited water.

Start by writing a question and developing a hypothesis. Write out how you plan to conduct the investigation and identify the variables that need to be controlled. Predict what your results will show.

INVESTIGATION 2: DESCRIBING WATER TRANSPORT IN A PLANT

AIM

To trace the path of water through a celery stalk to the leaves

YOU NEED

- ☑ fresh celery stalk that has been standing in a beaker of water and red dye for a few hours
- ☑ single-edge razor blade
- ☑ microscope glass slides and coverslips
- ☑ dropper bottle containing water
- ☑ tile or cutting board
- ☑ microscope

Warning

The razor blade, microscope glass slides and coverslips are sharp. Take care as you handle them. If you cut yourself, tell your teacher immediately.

Carry the microscope with two hands and always use it on a level surface.

WHAT TO DO

1 Write your own question related to the aim.

2 Remove the celery from the dye solution and examine the stalk and leaves carefully. Try to observe the path the dye has taken up the stalk and into the leaves. Draw a labelled diagram to show this.

3 On the tile or cutting board, cut thin sections across the stalk with the razor blade. Place the thinnest section on the microscope slide with a drop of water and cover it with a coverslip.

4 Examine the sections under low magnification and draw a diagram showing the distribution of red colour.

5 If you have time and another piece of celery, cut vertically up the stalk about halfway, and immerse one side of the stalk in red dye and the other in blue dye. Repeat steps 2–4.

WHAT DO YOU THINK?

1 Describe how the dye is distributed in the stalk. Is the dye found in certain places or is it throughout the stalk?

2 What (if any) details could you see using the microscope that you could not see without it?

3 Identify the plant tissue in which the dye travelled up the stalk.

4 Describe the process by which the dye moved from the water in the beaker to the leaves.

CONCLUSION

Write a conclusion linked to the aim of this investigation. Did you answer the question you asked at the beginning?

3 REVIEW

REMEMBERING

1 Match the structure (a–d) to its function (i–iv).

Structure	Function
a Epidermis	i Transports water
b Guard cell	ii Protection; reduces water loss
c Xylem	iii Non-cellular protective layer
d Cuticle	iv Controls opening of stoma

2 **List** four requirements of plants.

3 **State** the word equation for photosynthesis.

UNDERSTANDING

4 **Explain** three ways in which plants have adapted to recover after a bushfire.

5 **Name** the essential nutrient that carnivorous plants are lacking. Explain how they make up for this deficit.

6 The image below is a cross-section of a plant stem. The pointers labelled A–D indicate different tissues.
   a **Name** the type of tissues labelled A and B.
   b Which letter is pointing to tissue cells that contain chloroplasts?

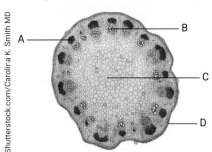

APPLYING

7 a **Describe** two examples of practices used by First Nations Australians that demonstrate sustainable harvesting.
   b **Explain** how these practices demonstrate sustainability.

8 **Predict** the effect on water loss if a plant was sealed in a plastic bag.

9 Grasses are often planted in areas subject to strong winds and high erosion. **Explain** this practice.

10 **Describe** the pathway of a water molecule from the soil, through a plant, to the atmosphere.

11 Variegated plants are plants with green and white stripes on their leaves. Would you expect these plants to grow as fast as non-variegated plants? Give reasons for your answer.

12 **Predict** the effect of blocking the stomata on the rates of photosynthesis and transpiration.

13 How could you decide whether sap from a plant was from the phloem or from the xylem?

14 **Explain** why a plant loses weight if it is kept in the dark.

ANALYSING

15 **Compare** and **contrast** the root system and the shoot system of a plant.

16 Carbon dioxide levels in the atmosphere are increasing. **Predict** the possible effect of this on the rate of photosynthesis of plants.

17 Where in a plant would you expect photosynthesis not to occur?

18 How does xylem differ from phloem in:
   a function?   b structure?

19 **Explain** how the root hair cells of plants and the villi of the small intestine of animals are similar/different.

EVALUATING

20 How important do you think phytoplankton are in terms of removing carbon dioxide from the atmosphere? Give reasons for your answer.

21 **Explain** the importance of stomata in the transpiration of water.

22 **Explain** two ways that the shoot system and root system work together to ensure a plant's survival.

CREATING

23 **Create** a story that names and describes the 'journey' of some carbon. The carbon begins as glucose in an apple. You eat the apple and breathe out the apple's carbon as carbon dioxide. This finally becomes part of an orange on a tree in your garden.

BIG SCIENCE CHALLENGE PROJECT

1 Connect what you have learned

In this chapter you have learned a lot about the structure and function of plant tissues and organs, how plants meet their needs and how they interact with their environment. Create a mind map to show how the information that you have learned is connected.

2 Check your thinking

Think about the two different plants in the images at the start and end of this chapter. What types of environments do they live in? How do they deal with those environments?

3 Make an action plan

Conduct some research into the different ways that plants are adapted to the environment in which they live. Choose two environments that are very different and choose one plant from each that you find interesting. This could be a cactus in the desert and a snow gum in an alpine region. Think about what you have learned in this chapter so that you can understand the challenges faced by the plants in each environment. Compare how the plants can deal with these environmental challenges.

4 Communicate

Use your knowledge, understanding and research to create a poster to communicate your findings to your classmates.

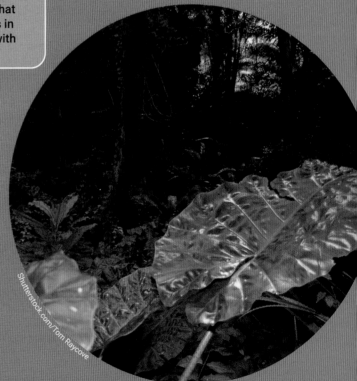

Shutterstock.com/Tom Raycove

Classifying matter

4.1 Atoms and elements (p. 86)

Atoms are the building blocks of all matter.

4.2 The changing concept of the atom (p. 90)

Advances in science over time have developed and refined concepts of the atom.

4.3 The periodic table (p. 94)

The periodic table is an important and relevant tool in chemistry.

4.4 Molecules and compounds (p. 98)

At the atomic level, substances can be classified as elements, compounds or molecules.

4.5 Models and chemical formulas (p. 102)

Elements and compounds can be represented in a variety of ways.

4.6 Mixtures: solutions, suspensions and colloids (p. 104)

Mixtures exist in several forms.

4.7 SCIENCE AS A HUMAN ENDEAVOUR: Chemistry allows for drug testing in sport (p. 108)

Chemical testing for the use of banned substances in sport has improved with advances in technology.

4.8 SCIENCE INVESTIGATIONS: Using an electronic scale (p. 109)

1 Density of common metals

2 3D molecular models

State Library of South Australia, B 42723

Salt is part of written history as far back as 4700 years ago. Historically, cities and towns have been created or destroyed because of salt production. Centuries ago, the French government used to force citizens to buy salt from a royal depot with a high salt tax. This unpopular tax is said to have been one of the causes of the French Revolution.

In centuries past, before refrigeration, salt was used to preserve food.

We need some salt in our diet, but too much can have detrimental effects, such as high blood pressure, which can lead to heart disease and kidney problems.

▶ **How can such a simple substance have such profound effects?**

#4 SCIENCE CHALLENGE ACCEPTED!

At the end of this chapter, you can complete Big Science Challenge Project #4. You can use the information you learn in this chapter to complete the project.

Assessments
- Prior knowledge quiz
- Chapter review questions
- End-of-chapter test
- Portfolio assessment task: Science investigation

Videos
- Science skills in a minute: Collecting and organising data **(4.8)**
- Video activities: What is an atom? **(4.1)**; The discovery of the atom **(4.2)**; Introduction to the periodic table **(4.3)**; Elements and compounds **(4.4)**; Solutions **(4.6)**; Testing athletes for doping **(4.7)**

Science skills resources
- Science skills in practice: Collecting and organising data **(4.8)**
- Extra science investigations: Observing elements and compounds **(4.4)**; Modelling elements and compounds **(4.5)**

Interactive resources
- Simulation: Build a molecule **(4.5)**; Concentration of solutions **(4.6)**
- Label: Molecule or compound? **(4.4)**
- Drag and drop: Properties of elements **(4.3)**; Solutions, suspensions and colloids **(4.6)**

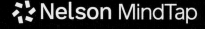 **Nelson MindTap**

To access these resources and many more, visit:
cengage.com.au/nelsonmindtap

4.1 Atoms and elements

BY THE END OF THIS MODULE, YOU WILL BE ABLE TO:

✓ explain how atoms are the building blocks of all matter

✓ distinguish between an atom and an element

✓ recognise that atoms and elements can be represented by names, symbols and diagrams

✓ recall the three categories used to classify all elements.

Video activity
What is an atom?

Quiz
Elements

GET THINKING

In this module you will learn that the building blocks of all matter are tiny particles called atoms. How do we know there are many different atoms?

Atoms: the building blocks of all matter

Imagine a majestic snow-capped mountain. From a distance the snow looks like it is one big piece, but up close we know it is made up of many trillions of snowflakes. Likewise, we know a sandy beach is made up of billions and billions of grains of sand. Whether we refer to individual snowflakes or sand grains, they are both the building blocks of something bigger.

Similarly, our lives consist of a wide variety of objects, built from smaller things, made from many different materials, but they all have one thing in common. If you were able to cut any object in half, halving it in size, over and over, millions of times, you would eventually get the basic piece of matter that can no longer be divided – that is called an **atom**.

atom
the smallest part of an element that contains the properties of that element

The word 'atom' comes from the Greek term *atomos*, meaning 'indivisible'. Atoms are much too small to see with the human eye. One single grain of sand is actually a large collection of atoms – billions of billions of them!

▲ **FIGURE 4.1.1** The metal copper is a shiny orange-brown solid made up of copper atoms.

Why are we all different?

If everything in the universe is made of the same thing – atoms – then why doesn't everything look the same? The answer is that there are 92 different atoms that make up everything you've ever seen. The properties of any object depend on their atoms and how they are connected.

For example, the metals copper and zinc are quite different. Copper (Figure 4.1.1) is a shiny orange-brown solid at room temperature. Zinc (Figure 4.1.2) is a shiny grey solid at room temperature. These two metals look different because copper is made up of copper atoms and zinc is made up of zinc atoms.

▲ **FIGURE 4.1.2** The metal zinc is a shiny grey solid made up of zinc atoms.

What do atoms look like?

It is hard to imagine what atoms look like because they are so small. The diagrams in Figure 4.1.3 represent what three different pure substances would look like if we could magnify them enough to see how the atoms would look and how they were arranged. The individual atoms are represented by the small, coloured spheres. Each pure substance shown has different coloured spheres (atoms aren't individually coloured, but scientists represent them this way to tell them apart). This shows that the atoms of each pure substance are different from each other and so the substances themselves must also be different. Diagrams such as these are one way to visualise the type and arrangement of atoms in different substances.

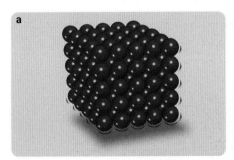

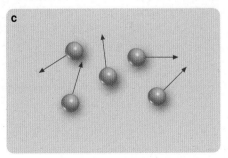

▲ **FIGURE 4.1.3** **(a)** A solid pure substance made up of one type of atom; **(b)** a liquid pure substance made up of one type of atom; **(c)** a gaseous pure substance made up of one type of atom

An element contains identical atoms

The atoms represented as red in Figure 4.1.3a are all the same colour and size. This is also true for the blue and yellow atoms. This means all three substances are pure; each of them is made up of only one type of atom.

The illustrations represent a type of pure substance known as an **element**. An element is a pure substance that is made up of only one type of atom. Elements cannot be broken down into other simpler substances because they already exist in their most basic form. Each element has a name and is represented by a chemical symbol. The symbol for each element often comes from the element's name. For example, the element chlorine is represented by the symbol Cl.

The metal copper, represented by the symbol Cu, is an element because it is made up only of copper atoms. If it were possible to cut a piece of copper in half, halving the size, over and over again until it was down to the individual atoms, they would still be copper atoms and therefore it would still be copper. The metal zinc is represented by the symbol Zn.

Today, scientists recognise 118 different elements. The first 92 elements occur naturally on Earth, while the last 26 were created by scientists, and many only existed in the laboratory for fractions of a second. These human-made or synthesised elements are called **transuranic**.

Elements, much like the atoms they are made of, have their own set of characteristics, which scientists call **properties**. Examples of properties are colour, melting point and density. It is these properties that allow elements to be classified, or grouped, into three categories: **metals**, **non-metals** and **metalloids**. This is shown in Figure 4.1.4.

Approximately 80 per cent of the known elements are metals, and they all have similar properties. We will learn more about these metallic properties in Module 4.3.

element
a pure substance made up of only one type of atom; cannot be broken down into a simpler substance

transuranic
an element with an atomic number greater than that of uranium

property
a characteristic or feature of a substance

metal
a chemical element that has certain properties, such as conducting heat and electricity, being malleable and being ductile

non-metal
a chemical element that has certain properties, such as being brittle, having a non-shiny appearance and low melting point

metalloid
a chemical element that has properties in between those of a metal and a non-metal

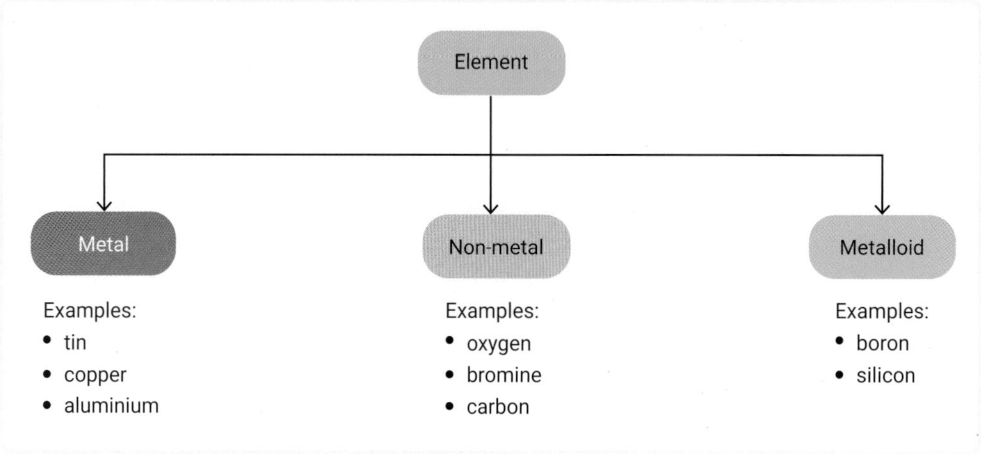

▲ FIGURE 4.1.4   The classification of elements

Rethinking what you know about elemental metals

If you ate cereal for breakfast today, it is likely that it was enriched with nutrients such as iron. Believe it or not, the iron you ate is the same as the iron used every day in buildings, automobiles and electronics. Iron, calcium, sodium (found in table salt) and potassium are all examples of metallic elements with important roles in industry as well as in human health.

4.1 LEARNING CHECK

1 **Recall** what all matter is made up of.

2 **Identify** the names of the elements represented by the following symbols.

   a Cl

   b Cu

   c Zn

3 All elements are classified into three categories. **List** them.

4 **Create** a relevant question and answer for each of the words below. Make sure the word is included.

   a Property

   b Element

   c Metalloid

5 **Explain** what is meant by the phrase 'atoms are the building blocks of all matter'.

6 Refer to Figure 4.1.5 to answer the following questions.

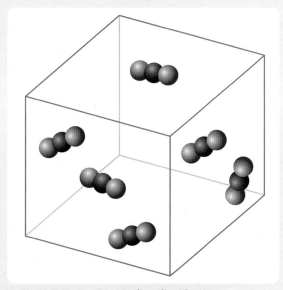

▲ FIGURE 4.1.5   Pure carbon dioxide gas

   a **Determine** the total number of atoms represented here.

   b **Explain** what the different colours represent.

7 **Analyse** Figure 4.1.4 to give other examples of metals, non-metals and metalloids.

8 **Compare** an atom and an element, discussing their similarities and differences.

4.2 The changing concept of the atom

BY THE END OF THIS MODULE, YOU WILL BE ABLE TO:

✓ explain how ideas in science change over time

✓ describe how Democritus, Antoine Lavoisier and John Dalton each contributed to the changing concept of the element.

Video activity
The discovery of the atom

GET THINKING

Thousands of years ago, different philosophies and cultures used what they called the 'classical elements' to explain the nature of the world around them. The four main classical elements were earth, air, water and fire. How has our understanding of the nature of matter changed since then?

▲ **FIGURE 4.2.1**
Democritus' (460 BCE) model of the atom

A philosophical question

A renowned Greek philosopher, Democritus, was having a philosophical debate approximately 2500 years ago. In 460 BCE, he argued that a change in appearance – for example, dough becoming a loaf of bread – is not something coming from nothing. As such, he proposed that changes in appearance were merely very small indivisible things – atoms – rearranging themselves.

The atomic ideas of Democritus (Figure 4.2.1) were not accepted, and largely ignored, until much later. Because atoms are too small to see, and hence there was no proof of their existence, the ideas were too theoretical for the time.

Greek philosopher who proposed that all matter was made of small units named atoms.

Democritus
(460 BCE)

English scientist who published a book, *New System of Chemical Philosophy,* to explain his atomic theories. Many of his ideas are still accepted today.

John Dalton
(1803)

Antoine Lavoisier
(1789)

French chemist who wrote the first 'modern' chemistry textbook, entitled *Elements of Chemistry,* in which he proposed that an element should be defined as a substance that cannot be broken up into any simpler substances.

▲ **FIGURE 4.2.2** Timeline showing the development of atomic understanding

Advances in science occur over time

How did the ancient ideas about atoms and elements develop into our current understanding? As with all scientific development, advances in knowledge mostly arise through scientific research. When scientists investigate a hypothesis (a proposed explanation), they are trying to expand their knowledge of that idea. Sometimes the results from their experiments support their understanding. At other times, new evidence will suggest that their understanding needs to be modified. For example, the 'Get thinking' section of this module states that, thousands of years ago, philosophers held a common belief that everything in the natural world was made up of four 'elements': earth, air, water and fire. Between then and now, we've come to understand, through scientific research, that everything is made of 92 different naturally occurring elements. This is how scientific theories are refined and improved; often slowly. With continued advances in technology, scientists are able to develop new techniques to find out more about atoms and elements.

Figure 4.2.2 is a timeline of the development of atomic understanding. Let's take a closer look at the developing concept of an element from Democritus (460 BCE) to John Dalton (1803). The refinement of an element continues past Dalton; you will learn more about that later.

J.J. Thompson
(1897)

Niels Bohr
(1913)

Dmitri Mendeleev
(1869)

Ernest Rutherford
(1909)

Erwin Schrödinger
(1926)

Russian chemist who created the first periodic table of elements, arranged by increasing atomic weight.

Chapter 4 | Classifying matter   **91**

Developing and refining the concept of an element

It was not until the 1700s, more than 2000 years after Democritus' original concept, that the idea of matter being composed of particles re-emerged. By this time the scientific method was established, and chemists were able to support new ideas with existing evidence.

▲ FIGURE 4.2.3   Antoine Lavoisier

French chemist Antoine Lavoisier (1743–94) (Figure 4.2.3) wrote what many consider to be the first 'modern' chemistry textbook, entitled *Elements of Chemistry*, in which he proposed that an element should be defined as a substance that cannot be broken up into any simpler substances. In his book he listed the 23 elements known at that time. Thus, the term 'element', as we currently use it today, was defined.

Soon after, English teacher and scientist John Dalton (1766–1844) (Figure 4.2.4) examined the information accumulated by many chemists, including Lavoisier, as well as the results of his own experiments. Being a scientist, Dalton tried to find a logical way to explain existing principles. His conclusion was startlingly simple. The early Greek philosophers, such as Democritus, had been correct. Even though atoms cannot be seen, all matter must be made of separate, tiny particles! Dalton believed atoms must be solid and indestructible, much like billiard balls (Figure 4.2.5). This type of representation of an idea is known as a **scientific model**. Scientific models can exist in many forms (e.g. mathematical and computer-based) and are used to make a particular part of a system or idea easier to understand, visualise or define.

scientific model
a physical, conceptual or mathematical representation of a real phenomenon that is difficult to observe directly

▲ FIGURE 4.2.4   John Dalton

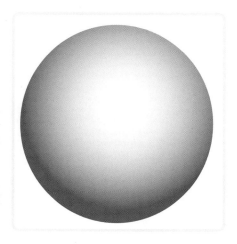

▲ FIGURE 4.2.5   John Dalton's model of the atom was similar to that of Democritus.

9780170463027

In 1808, John Dalton published a book, *A New System of Chemical Philosophy*, to explain his theory (Figure 4.2.6). In advocating his atomic theory, he also recommended symbols be used to represent different atoms.

Although the model of the atom has continued to evolve, the main ideas of John Dalton's theory that have been accepted by today's scientists are shown in Figure 4.2.7.

ELEMENTS
Hydrogen
Azote
Carbon
Oxygen
Phosphorus
Sulphur
Magnesia
Lime
Soda
Potash
Strontian
Barytes
Iron
Zinc
Copper
Lead
Silver
Gold
Platina
Mercury

Alamy Stock Photo/Science History Images

▲ **FIGURE 4.2.6** A page from Dalton's *A New System of Chemical Philosophy* (1808)

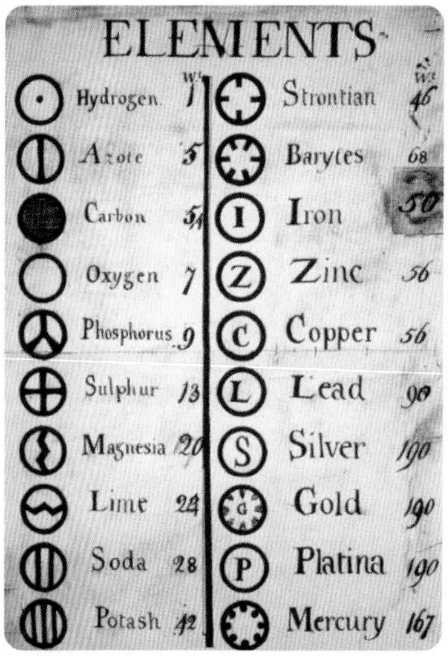

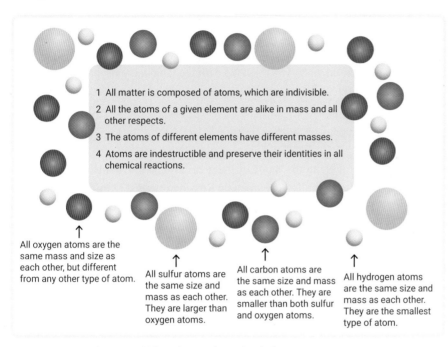

1 All matter is composed of atoms, which are indivisible.
2 All the atoms of a given element are alike in mass and all other respects.
3 The atoms of different elements have different masses.
4 Atoms are indestructible and preserve their identities in all chemical reactions.
All oxygen atoms are the same mass and size as each other, but different from any other type of atom.
All sulfur atoms are the same size and mass as each other. They are larger than oxygen atoms.
All carbon atoms are the same size and mass as each other. They are smaller than both sulfur and oxygen atoms.
All hydrogen atoms are the same size and mass as each other. They are the smallest type of atom.

▲ **FIGURE 4.2.7** Accepted ideas from John Dalton's theory

4.2 LEARNING CHECK

1 **List** Dalton's four main ideas of atomic theory that are still accepted today.
2 **Identify** two factors that cause scientific ideas to change over time.
3 **Explain** the importance of a scientific model.
4 Use text and diagrams to show how you would **explain** the idea of Dalton's atom to a primary school student.
5 **Create** your own timeline, detailing how the concept of an element changed from Democritus to Antoine Lavoisier to John Dalton.
6 **Discuss** (giving reasons for and against) why scientists should work together when developing new scientific theories.

Video activity
Introduction to the periodic table

Interactive resource
Drag and drop: Properties of elements

GET THINKING

The periodic table, first developed in 1869, consolidates much of the existing knowledge about chemistry onto a single page. Do you think the periodic table remains an important and relevant tool in chemistry?

The original periodic table

▲ FIGURE 4.3.1 Dmitri Mendeleev published the first periodic table.

The **periodic table** of elements is one of the great intellectual achievements of humankind. It was originally developed by Russian chemist Dmitri Mendeleev (1834–1907) (Figure 4.3.1). It catalogues all the elements in the known universe. The arrangement of elements in the periodic table also provides patterns and clues about their properties.

In 1869, Mendeleev discovered that by grouping the 63 known elements with regard to their properties (columns) and generally from lightest to heaviest (rows), there were patterns, or periodicity (thus, periodic table), within the rows and columns of the table (Figure 4.3.2). He also proposed the **densities** and weights of the elements, known as **atomic weights**.

Mendeleev made some incorrect predictions. This was not surprising considering that only just over half of the elements had been discovered when he was formulating his table. Later discoveries showed that some of the elements had

periodic table
a method of arranging elements by increasing atomic number **density** the mass of a substance in a specific volume

atomic weight
the mass of an atom on Earth

Tabelle II.

Reihen	Gruppe I. — R²O	Gruppe II. — RO	Gruppe III. — R²O³	Gruppe IV. RH⁴ RO²	Gruppe V. RH³ R²O⁵	Gruppe VI. RH² RO³	Gruppe VII. RH R²O⁷	Gruppe VIII. — RO⁴
1	H=1							
2	Li=7	Be=9,4	B=11	C=12	N=14	O=16	F=19	
3	Na=23	Mg=24	Al=27,3	Si=28	P=31	S=32	Cl=35,5	
4	K=39	Ca=40	—=44	Ti=48	V=51	Cr=52	Mn=55	Fe=56, Co=59, Ni=59, Cu=63.
5	(Cu=63)	Zn=65	—=68	—=72	As=75	Se=78	Br=80	
6	Rb=85	Sr=87	?Yt=88	Zr=90	Nb=94	Mo=96	—=100	Ru=104, Rh=104, Pd=106, Ag=108.
7	(Ag=108)	Cd=112	In=113	Sn=118	Sb=122	Te=125	J=127	
8	Cs=133	Ba=137	?Di=138	?Ce=140	—	—	—	— — — —
9	(—)	—	—	—	—	—	—	
10	—	—	?Er=178	?La=180	Ta=182	W=184	—	Os=195, Ir=197, Pt=198, Au=199.
11	(Au=199)	Hg=200	Tl=204	Pb=207	Bi=208		—	
12	—	—		Th=231	—	U=240	—	— — — —

der chemischen Elemente.

▲ FIGURE 4.3.2 Dmitri Mendeleev's periodic table

been placed in the wrong columns and rows. However, Mendeleev did the important work of establishing the basic structure of the periodic table.

The periodic table we see and use today is not the same as the one originally developed by Dmitri Mendeleev. It has changed over time as scientists have discovered new elements and developed a greater understanding of the atoms that make up elements.

One of the most important things Mendeleev realised was that he needed to leave spaces for elements that at the time were still undiscovered. Scientists could work out the properties of these unknown elements by looking at the properties of the surrounding elements. Later, when some of these elements were discovered, Mendeleev's brilliance was recognised and is still acknowledged today. He is considered the 'father of the periodic table'.

The modern periodic table

While Mendeleev organised his periodic table by increasing weight of elements, the modern periodic table (Figure 4.3.3, page 96) organises elements by increasing atomic number. (You will learn more about the atomic number in Year 9.)

The periodic table provides the name and **chemical symbol** of every known element. Symbols are used as abbreviations for elements, which is quicker and easier than writing the full word. The use of chemical symbols is a type of internationally recognised 'shorthand' that has no language barrier – the chemical symbol for each element is the same worldwide.

chemical symbol
a letter or letters of the Latin alphabet used to represent an atom of a specific element

The first letter of the chemical symbol is always a capital. If there is a second letter, it is always lower case. Many of the symbols are the first letter of the name of the element, such as H for hydrogen and C for carbon. Other elements have two letters in their symbols, such as Ca for calcium and Co for cobalt. This is so the elements carbon, calcium and cobalt are not confused. The chemical symbols of some other elements seem to have no relationship to their names, such as Fe for iron and Sn for tin. This is because the names of these elements come from another language. The Latin word for iron is *ferrum* and for tin it is *stannum*.

The modern periodic table is an unusual shape. This is because the elements on the periodic table are not randomly arranged. The elements are arranged into numbered (1–7) rows, called **periods**, according to their increasing atomic number. The elements are also sorted into numbered (1–18) columns, called **groups**, according to how similar their chemical properties are.

period
a horizontal row in the periodic table

group
a vertical column in the periodic table

For example, locate group 1 in Figure 4.3.3. It contains, from top to bottom, the following elements: H (hydrogen), Li (lithium), Na (sodium), K (potassium), Rb (rubidium), Cs (caesium) and Fr (francium). Note that element names, when written, are not proper nouns and do not need to be capitalised. The elements in this group have similar chemical properties: they are very reactive (explosive), especially in water, and they have low melting and boiling points. They are also less dense than most metals, a physical property they have in common.

The periodic table is often colour-coded to show the three classifications of all elements discussed in Module 4.1: metals, non-metals and metalloids. The properties of each element determine where the element falls within these three categories. The general properties of elements in each of the three classifications – metals, non-metals and metalloids – are shown in Table 4.3.1.

Group 1

| 1 H hydrogen | | | | | | | | | | | | | | | | | 18 He helium |

Key: atomic number / Symbol / element name

State at room temperature: Solid | Liquid | Gas | Synthetic

Classification: Metal | Non-metal | Metalloid

Periodic table (main body):

Period	Group 1	2	3	4	5	6	7	8	9	10	11	12	13	14	15	16	17	18
1	1 H hydrogen																	2 He helium
2	3 Li lithium	4 Be beryllium											5 B boron	6 C carbon	7 N nitrogen	8 O oxygen	9 F fluorine	10 Ne neon
3	11 Na sodium	12 Mg magnesium											13 Al aluminium	14 Si silicon	15 P phosphorus	16 S sulfur	17 Cl chlorine	18 Ar argon
4	19 K potassium	20 Ca calcium	21 Sc scandium	22 Ti titanium	23 V vanadium	24 Cr chromium	25 Mn manganese	26 Fe iron	27 Co cobalt	28 Ni nickel	29 Cu copper	30 Zn zinc	31 Ga gallium	32 Ge germanium	33 As arsenic	34 Se selenium	35 Br bromine	36 Kr krypton
5	37 Rb rubidium	38 Sr strontium	39 Y yttrium	40 Zr zirconium	41 Nb niobium	42 Mo molybdenum	43 Tc technetium	44 Ru ruthenium	45 Rh rhodium	46 Pd palladium	47 Ag silver	48 Cd cadmium	49 In indium	50 Sn tin	51 Sb antimony	52 Te tellurium	53 I iodine	54 Xe xenon
6	55 Cs caesium	56 Ba barium		72 Hf hafnium	73 Ta tantalum	74 W tungsten	75 Re rhenium	76 Os osmium	77 Ir iridium	78 Pt platinum	79 Au gold	80 Hg mercury	81 Tl thallium	82 Pb lead	83 Bi bismuth	84 Po polonium	85 At astatine	86 Rn radon
7	87 Fr francium	88 Ra radium		104 Rf rutherfordium	105 Db dubnium	106 Sg seaborgium	107 Bh bohrium	108 Hs hassium	109 Mt meitnerium	110 Ds darmstadtium	111 Rg roentgenium	112 Cn copernicium	113 Nh nihonium	114 Fl flerovium	115 Mc moscovium	116 Lv livermorium	117 Ts tennessine	118 Og oganesson

Lanthanides and actinides:

57 La lanthanum	58 Ce cerium	59 Pr praseodymium	60 Nd neodymium	61 Pm promethium	62 Sm samarium	63 Eu europium	64 Gd gadolinium	65 Tb terbium	66 Dy dysprosium	67 Ho holmium	68 Er erbium	69 Tm thulium	70 Yb ytterbium	71 Lu lutetium
89 Ac actinium	90 Th thorium	91 Pa protactinium	92 U uranium	93 Np neptunium	94 Pu plutonium	95 Am americium	96 Cm curium	97 Bk berkelium	98 Cf californium	99 Es einsteinium	100 Fm fermium	101 Md mendelevium	102 No nobelium	103 Lr lawrencium

▲ **FIGURE 4.3.3** The modern periodic table (current as of October 2022).

▼ **TABLE 4.3.1** Classifications of all elements on the periodic table, and their general properties

	Metals	Non-metals	Metalloids
Physical properties	• Shiny metallic lustre • Solid at room temperature • Good conductors of heat and electricity • Malleable • Ductile • High melting points	• Dull • Poor conductors of heat and electricity • Brittle • Low melting point	• Have some properties of both metals and non-metals
Examples	iron aluminium gold	chlorine sulfur carbon	boron germanium arsenic

9780170463027

1 **Recall** what a period and a group are on the periodic table.

2 **Identify** the period number of the following elements.

   a Boron

   b Helium

   c Potassium

3 **Identify** the group number of the following elements.

   a Strontium

   b Phosphorus

   c Tin

4 Each element on the periodic table has a unique chemical symbol. **Explain** why this is important.

5 **Contrast** the organisation of Mendeleev's periodic table to the current organisation of the periodic table.

6 Search the Internet to find an image of each of the metals aluminium and zinc. **Explain** why they are not the same element even though they are in the same state at room temperature and are the same colour. Research the physical properties of aluminium and zinc that are different from one another. Use the word 'atom' in your answer.

7 Match the elemental properties (a–f) to their correct chemical name and symbol (i–vi). Use the information in Table 4.3.1. You may also need to do some extra research.

Elemental properties	Chemical name and symbol
a I am shiny and gold in colour. I conduct heat and electricity and am malleable. I am prized for my beauty.	i chlorine, Cl
b I am a gas, yellow in colour. I do not conduct electricity. I am commonly used in swimming pools as a disinfectant.	ii potassium, K
c I am shiny silver in colour and am one of the lighter elements of my kind. I conduct electricity and am malleable. I am used in soft drink cans and can be recycled.	iii gold, Au
d I am a colourless gas that makes up most of the air you breathe, but I am not oxygen.	iv neon, Ne
e I am soft and white with a silvery lustre. I conduct electricity and am malleable. I ignite with a purple flame when mixed with water.	v aluminium, Al
f I am a colourless, odourless inert gas but, when in a vacuum, I change to reddish-orange. Because of this I am often used in colourful electric signs.	vi nitrogen, N

8 Reflect on why Dmitri Mendeleev has been given the title 'the father of the periodic table'.

4.4 Molecules and compounds

BY THE END OF THIS MODULE, YOU WILL BE ABLE TO:

✓ distinguish between elements, compounds and molecules

✓ list known monatomic and diatomic elements.

Video activity
Elements and compounds

Interactive resource
Label: Molecule or compound?

Extra science investigation
Observing elements and compounds

molecule
the smallest particle of a substance that is capable of separate existence

chemical bond
a force that holds atoms together

molecule of an element
a molecule in which two or more atoms of the same non-metal element are chemically bonded together

molecule of a compound
a molecule in which two or more atoms of different non-metal elements are chemically bonded together; also known as a molecular compound

GET THINKING

A large part of chemistry relies on the study of atoms connecting, or chemically bonding. Think of one of your favourite objects. Whether it is an electronic device or something simple such as a netball, it was probably made by combining elements.

What is a molecule?

Water is a common **molecule**. Water is made of two hydrogen atoms and one oxygen atom: H_2O. Looking at a periodic table (Figure 4.3.3, p. 96), what do you notice about the atoms in a water molecule? Both hydrogen and oxygen are non-metals. Therefore, a molecule is a type of particle where two or more non-metal atoms are held together by a **chemical bond**.

The molecules shown in Table 4.4.1 represent two types of molecules. If all the atoms in the molecule are the same – such as ozone, O_3, which is three oxygen atoms – then it is called a **molecule of an element**. If two or more different types of non-metal atoms are bonded together – such as water, H_2O – then it is called a **molecule of a compound**, or, more commonly, a molecular compound. In a nutshell, for a substance to claim fame as a molecule, the general rule is that it is made entirely of non-metals!

▼ **TABLE 4.4.1** Comparing a molecule of an element to a molecule of a compound

	Molecule of an element	Molecule of a compound
Types of atoms	• Atoms are identical. • Atoms are non-metals.	• Atoms are different. • Atoms are non-metals.
3D model	Ozone, O_3	Water, H_2O
2D model	Ozone, O_3	Water, H_2O

9780170463027

What is a compound?

A **compound** is a substance made up of two or more different types of atoms chemically bonded together. Similar to molecules, there are two types of compounds, and, generally, the type is determined by the bond between the atoms, and whether the bonded atoms are metals or non-metals.

Molecular compound

The first type of compound is a **molecular compound** – a compound formed between two different types of non-metal atoms. Water is a molecular compound made up of the non-metals hydrogen and oxygen.

Ionic compound

The second type of compound is an **ionic compound** – a compound formed when the two different atoms are a metal and a non-metal. Table salt (sodium chloride, NaCl), is an example of an ionic compound made up of the metal sodium and non-metal chlorine.

compound
a chemical substance made up of two or more different atoms chemically bonded together; bonded atoms can be a metal and a non-metal (ionic compound), or two non-metals (molecular compound)

molecular compound
a compound formed when two different non-metals chemically bond

ionic compound
a compound formed when a metal and a non-metal chemically bond

▼ TABLE 4.4.2 Comparing a molecular compound to an ionic compound

	Molecular compound	Ionic compound
Types of atoms	• Atoms are different. • Atoms are non-metals.	• Atoms are different. • Atoms are metals and non-metals.
3D model	Water, H_2O	Salt, NaCl
2D model	H H O Water, H_2O	Salt, NaCl

Molecule or compound?

Sometimes a visual representation, or a model, can be useful in understanding an idea. Figure 4.4.1 shows how compounds and molecules relate to one another in a Venn diagram. This figure shows water, H_2O, in the overlapping area, which means it is both a molecule and a compound. Ozone, O_3, is only a molecule. Salt, NaCl, is only a compound. More specifically, we learned it is an ionic compound, since it is made of a metal (sodium, Na) bonded to a non-metal (chlorine, Cl).

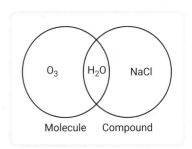

▲ FIGURE 4.4.1 The relationship between compounds and molecules

Diatomic and monatomic elements

Figure 4.4.2 shows a molecule of oxygen gas. Single oxygen atoms do not exist for very long on their own in nature. Instead, they 'pair up'. A molecule of oxygen gas is always made up of two oxygen atoms chemically bonded together. Oxygen is therefore **diatomic**, meaning it consists of two atoms. The chemical formula for oxygen gas is O_2, where the subscript '2' means there are two oxygen atoms. Note that every molecule that consists of two atoms is diatomic.

Diatomic elements include oxygen (O_2), iodine (I_2), bromine (Br_2), chlorine (Cl_2), fluorine (F_2), nitrogen (N_2) and hydrogen (H_2).

Can you think of a mnemonic phrase to help you remember the diatomic elements?

I	→	**I**odine, I_2
Bring	→	**B**romine, Br_2
Cookies	→	**C**hlorine, Cl_2
For	→	**F**luorine, F_2
Our	→	**O**xygen, O_2
New	→	**N**itrogen, N_2
Home	→	**H**ydrogen, H_2

Rarely, some atoms exist on their own in their natural state. These are known as **monatomic** (or **monoatomic**) elements. Only six of the 92 naturally occurring elements are monatomic: helium (He), neon (Ne), argon (Ar), krypton (Kr), xenon (Xe) and radon (Rn). Do you notice anything about the placement on the periodic table of all of these monatomic elements? They are all in the last group; they are known as noble gases because they are stable and unreactive.

diatomic
a molecule consisting of two atoms (*di*– means 'two'; *atomic* means 'atom')

▲ FIGURE 4.4.2
Oxygen, an example of a diatomic molecule and an element

monatomic/monoatomic
an element consisting of just one atom (*mono*– means 'one'; *atomic* means 'atom')

▲ FIGURE 4.4.3   Monatomic elements in everyday life: (a) neon lights; (b) helium balloons; (c) argon (in lightbulbs)

9780170463027

1 a Using the diagram below, begin at the Start (S square at top left) position and move
      to the End (E square at bottom right) position by identifying and alternating between
      diatomic and monatomic elements only.

S	H₂	Kr	N₂	Ne	Al	He	F₂	He	O₂	Po
Li	Be	Y	S	O₂	Ga	Br₂	Na	C	Rn	Sr
He	I₂	Ar	Cl₂	Rn	As	Rn	Rb	Sc	Br₂	Mn
Br₂	Te	K	Mg	Cs	Sn	N₂	Te	Ti	Kr	Cu
Xe	F₂	He	Br₂	Ne	O₂	He	Ba	Fe	H₂	E

   b Copy the frequency table into your notebook and complete the tally and total columns
      for monatomic and diatomic elements as you move through the maze.

      Frequency table: Monatomic and diatomic elements

Monatomic elements			
Symbol	**Atom this represents**	**Tally**	**Total**
Ar	argon		
He	helium		
Ne	neon		
Kr	krypton		
Rn	radon		
Xe	xenon		

Diatomic elements			
Symbol	**Atom this represents**	**Tally**	**Total**
Br₂	bromine		
Cl₂	chlorine		
F₂	fluorine		
H₂	hydrogen		
I₂	iodine		
N₂	nitrogen		
O₂	oxygen		

   c **Create** a bar chart to represent the data in your frequency table.

2 Identify whether or not the following arrangements exist. If they do exist, provide an
   example.

   a An atom of a compound

   b Molecules of elements

   c A molecular compound

   d Atoms of elements

3 **Explain** the difference between elements and compounds. Include a diagram in your
   answer.

BY THE END OF THIS MODULE, YOU WILL BE ABLE TO:
- ✓ create both 2D and 3D models of atoms, elements or compounds
- ✓ interpret chemical formulas for meaning.

Interactive resource
Simulation: Build a molecule

Extra science investigation
Modelling elements and compounds

chemical formula
a collection of symbols and numbers that represent the number of atoms in a molecule or compound

GET THINKING

Models and representations in science are needed to communicate ideas, inform, raise awareness and increase the sense of wonder in science. Think about ways that scientific knowledge is all around you – when you turn on the TV, cook food or even ride your bicycle.

Representations of elements and compounds

There are several ways that chemists represent elements and compounds, depending on the purpose or reason for that representation.

Chemical formulas

In Module 4.4, you saw diatomic oxygen written as O_2, where the '2' referred to two oxygen atoms. In instances where there are numerical subscripts in compounds or molecules, the letters representing the elements are no longer referred to as chemical symbols but are instead part of a **chemical formula**. Scientists use chemical formulas when there is more than one atom in a substance; the formula gives the elemental composition of molecules. It's similar to a recipe for cooking. The formula, or recipe, consists of the chemical symbol for the element, followed by a subscript that tells you the number of atoms present. Let's look at the formula for a molecule of carbon dioxide as an example (Figure 4.5.1).

The chemical formula is CO_2. There is no subscript for carbon, which means there is only one carbon atom. The subscript is written only if the number of atoms is greater than one. However, the oxygen has a '2' subscript, which means there are two oxygen atoms present. This recipe calls for one carbon and two oxygen atoms; the ratio of carbon atoms to oxygen atoms is 1:2. Some other examples are shown in Table 4.5.1.

3D models

We have seen that chemical formulas provide some information about the composition of the chemical substance. However, the chemical formula does not give information about the structure of the chemical substance – that is, how the atoms are joined together.

Your teacher may have several ball-and-stick model kits, which can be used as a continuation of the idea of atoms represented as spheres. These kits allow you to see compounds in three dimensions, which is how the compounds exist in the real world.

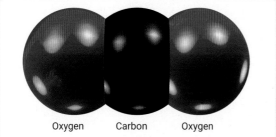

Oxygen   Carbon   Oxygen

The symbol for the element carbon. There is only one carbon atom in a molecule of carbon dioxide.

The symbol for the element oxygen

This number means there are two oxygen atoms in a molecule of carbon dioxide.

▲ **FIGURE 4.5.1** A carbon dioxide molecule and the chemical formula for carbon dioxide

▼ TABLE 4.5.1 The chemical formulas of some common chemicals

Name	Chemical formula	Ratio
water	H_2O	2 atoms of hydrogen and 1 atom of oxygen
carbon dioxide	CO_2	1 atom of carbon and 2 atoms of oxygen
oxygen	O_2	2 atoms of oxygen
methane	CH_4	1 atom of carbon and 4 atoms of hydrogen
ammonia	NH_3	1 atom of nitrogen and 3 atoms of hydrogen
sodium chloride	NaCl	1 atom of sodium and 1 atom of chlorine

The spheres, or balls, in a 3D model are meant to represent the atoms, and each type of atom is represented by a different colour. The connection, or bond, that holds the atoms together is represented by the stick. These models not only show the compounds as they are arranged in space (given that they are too small to see in real life), but they also provide hands-on experience for learners of chemistry. Study the ball-and-stick models of water and ethanol in Figure 4.5.2.

Structural formulas

A **structural formula** is a two-dimensional representation of a molecule that gives us information about which atoms are bonded to each other (Figure 4.5.3). Chemists also build 3D models of structural formulas, usually with a computer simulation. The 3D shape provides information about how compounds behave physically and chemically, which is particularly important in fields such as medicine.

Water, H_2O

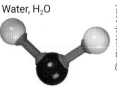

Ethanol, C_2H_5OH

▲ FIGURE 4.5.2
Ball-and-stick models of water and ethanol

Water, H_2O

Ethanol, C_2H_5OH

▲ FIGURE 4.5.3
Structural formulas of water and ethanol

structural formula
a graphic representation that shows the arrangement of atoms in a molecule or compound

4.5 LEARNING CHECK

1 **Explain** how subscripts are used to indicate the number of atoms present in a chemical formula.

2 **Explain** what is meant by the statement 'a chemical formula represents the ratio between atoms'.

3 **Compare** chemical formula to structural formula.

4 **Determine** the missing ratio or chemical formula for positions a–e in the table below. The first row has been done for you.

Name	Ratio of elements	Chemical formula
lithium carbonate	lithium : carbon : oxygen = 2:1:3	Li_2CO_3
phosphorus trichloride	phosphorus : chlorine = 1:3	a
silicon dioxide	silicon : oxygen = 1:2	b
nitrogen dioxide	c	NO_2
carbon tetrachloride	carbon : chlorine = 1:4	d
sulfur trioxide	e	SO_3

Chapter 4 | Classifying matter **103**

BY THE END OF THIS MODULE, YOU WILL BE ABLE TO:

✓ compare particle models of solutions, suspensions and colloids

✓ express amounts of substances in a solution as percentages.

Video activity
Solutions

Interactive resources
Drag and drop: Solutions, suspensions and colloids
Simulation: Concentration of solutions

---

mixture
a substance made up of two or more different types of particles, physically combined but not chemically bonded

GET THINKING

Air that you breathe is a mixture. The oceans you swim in are mixtures. Even the ground you stand on is a mixture. Mixtures are an important part of your daily life. What other mixtures can you think of?

Mixtures are everywhere!

Do you remember the scientific definition of a **mixture**? In science, a mixture is when two or more substances are physically combined, but not chemically bonded. In every room in your home, you will find examples of things that are mixtures. Your bookshelf holds a combination of different types of books, such as fiction and non-fiction. In the kitchen, you will find different things mixed in the fruit bowl and inside the refrigerator. Look on the label of your shampoo, toothpaste or deodorant, and you will see a list of chemical substances that have been combined to make that product. Your home is a mixture of mixtures!

Types of mixtures

Science – from the Latin word *scientia*, meaning 'knowledge' – is a collection of specific, organised knowledge. Studying mixtures is part of science, and there are several specific types of mixtures scientists have classified.

Solutions

When a sugar cube is added to a cup of tea and stirred, the small, white, solid grains seem to vanish (Figure 4.6.1). Where has the sugar cube gone? We know it hasn't 'gone' anywhere, because the tea now tastes sweet. When a substance seems to vanish like this, we say that the sugar has dissolved. The sugar is soluble in the hot water or tea.

▲ **FIGURE 4.6.1** The solute (sugar) dissolves in the solvent (tea) to form a solution.

The substance that dissolves is called the **solute**. The substance that does the dissolving is called the **solvent**. Together, the solute and the solvent form a special mixture known as a **solution**. In a sugar–water solution, sugar is the solute because it is the substance that dissolves, while the water is the solvent because it did the dissolving.

We can represent a solution as shown in Figure 4.6.2. This is known as a particle model. The yellow spheres represent sugar, which has a chemical formula of $C_{12}H_{22}O_{11}$. The solute (sugar in this case) enters the solvent (water in this case) initially as a whole cube. Upon dissolving, and becoming a mixed solution, the sugar molecules are evenly dispersed throughout the water. The sugar particles and the water particles are spread evenly throughout the entire solution, so that it is not possible to see the individual parts of the mixture. You can usually see through solutions, and you cannot see any solid parts in a solution.

solute
a substance in a solution that is dissolved

solvent
a substance that dissolves another substance to form a solution

solution
a mixture produced when a solute dissolves in a solvent

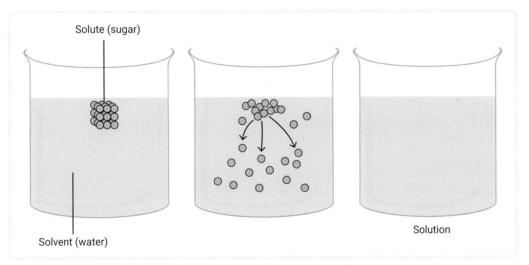

Solute (sugar)

Solvent (water)

Solution

▲ **FIGURE 4.6.2**   A solution is formed when a solute dissolves in a solvent.

In science, it is common practice to describe solution **concentration** in terms of percentages. Percentages in science are the same as the percentages you've learned about in maths – 'per cent' means 'parts out of 100'. Regarding solutions specifically, the amount (weight or volume) of a solute is expressed as a percentage of the total solution weight or volume.

concentration
the amount of solute present in a specified amount of solution

Let's look at a solution of cordial in water as an example. If you had a volume of 100 mL of cordial added to a volume of 400 mL of water, the total volume is:

    100 mL cordial
+   400 mL water
    _____
    500 mL total volume

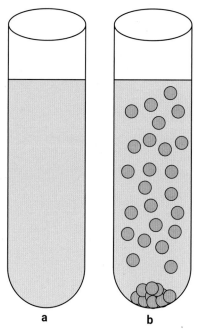

FIGURE 4.6.3 (a) A solution and (b) a suspension

So, the cordial is 100 'parts' out of a total 500 'parts'. To convert this fraction $\left(\frac{100}{500}\right)$ to a percentage, as is customary with solutions, follow the same steps you use in maths:

$$\% \text{ cordial solution} = \left(\frac{\text{volume of solute}}{\text{total volume of solution}}\right) \times 100$$

$$= \left(\frac{100 \text{ mL cordial}}{500 \text{ mL cordial}}\right) \times 100$$

$$= 0.20 \times 100$$

$$= 20\%$$

The cordial drink is a solution of 20% cordial in water.

Suspensions

When sand is mixed with water, the sand does not dissolve. For this reason, sand is said to be insoluble in water. A mixture is called a **suspension** when an insoluble solid is added to a liquid or a solution. Over time, the solid may settle to the bottom of the container, or it may float on the surface of the liquid or solution (Figure 4.6.3).

Colloids

suspension
a mixture of at least one insoluble solid and a liquid or a solution, where the insoluble substance settles to the bottom or rises to the top over time

colloid
a mixture of two or more insoluble substances that remains evenly mixed and does not settle over time

opaque
describes something that light cannot pass through

A **colloid** is another type of mixture. It results when two or more substances that are insoluble in one another are combined, but the insoluble particles of both substances are fairly small, so they are not heavy enough to settle the way they do in a suspension. Instead, they remain evenly spread throughout one another. The particles in a colloid are larger than in a solution, so the mixture is cloudy, not clear. Colloids may appear transparent, translucent or **opaque** depending on the concentration of the particles.

Ink is a colloid. It has solid pigment particles spread evenly in a liquid of oil or water. Ink is not a solution because the solid pigment particles do not dissolve in the oil or water. It is not a suspension either, because the mixture does not settle. Smoke is a colloid with small solid particles of ash mixed with air.

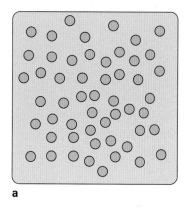
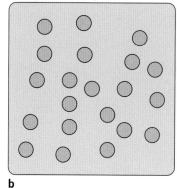
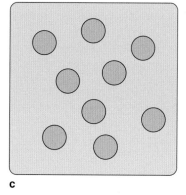

FIGURE 4.6.4 Particles within (a) solutions, (b) colloids and (c) suspensions vary in size.

9780170463027

1 Read the paragraph. **Explain** the meaning of each bolded word.

In my science class, my teacher did an experiment to talk about **mixtures**. The experiment, shown in the photo below, was called 'Elephant's toothpaste'. First, she put the **solute**, sodium iodide crystals, in a flask and dissolved them with the **solvent**, water. Next, she combined a **solution** of **30% hydrogen peroxide** with the sodium iodide solution and dish soap ... and, wow, there was a foaming eruption of something the teacher called an **opaque colloid**.

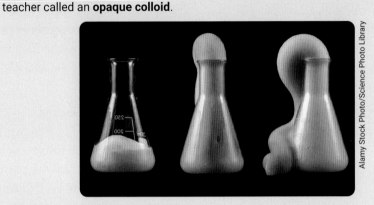

Alamy Stock Photo/Science Photo Library

2 **Calculate** the percentage of solute in each of these solutions.

a 100 mL solute per 900 mL solvent

b 25 mg solute dissolved in 100 mL solvent

3 Solutions are a type of mixture where concentration is often expressed as a percentage. The following solutions were mixed in the laboratory. **Determine** which solution has the highest percentage of solute.

Making a solution		
Beaker	**Amount of solute (mL)**	**Amount of solvent (mL)**
A	10	100
B	5	100
C	50	250
D	10	250

4 Draw a particle model for each of the following in your everyday life.

a A solution

b A suspension

c A colloid

4.7 Chemistry allows for drug testing in sport

BY THE END OF THIS MODULE, YOU WILL BE ABLE TO:

✓ explain how science plays a part in keeping sport fair.

Video activity
Testing athletes for doping

In sport, 'doping' is the use of banned substances such as drugs to improve performance. When an athlete deliberately or accidentally takes a performance-enhancing drug, a chemical signature is left behind in their body, sometimes remaining for weeks after exposure. Analytical chemistry is used to detect the chemical signature of banned substances. Often a very small amount of a broken-down banned substance remains in the athlete's bio-fluids (blood, urine or saliva).

In Module 4.2, we looked at how science ideas change and improve over time. Chemical testing for banned substances has improved with advances in technology, and refinements in chemical tests show a higher number of athletes with positive doping results. However, even though technology for testing is constantly developing and improving, designer drugs continue to be developed to escape or defy current testing procedures.

In Australia, an accredited body, Sport Integrity Australia, is responsible for testing athletes for an anti-doping rule violation. Sport Integrity Australia complies with the World Anti-Doping Code. This code regulates anti-doping policies across all sports and all countries in the world.

Because drugs are continually being developed to evade current tests, Sports Integrity Australia stores blood and urine samples from athletes for up to 10 years. These samples can be retested when new technology has been developed to detect new drugs. Science must respond by investing in research and development of technology and testing procedures to detect doping cheats and keep fairness in sport.

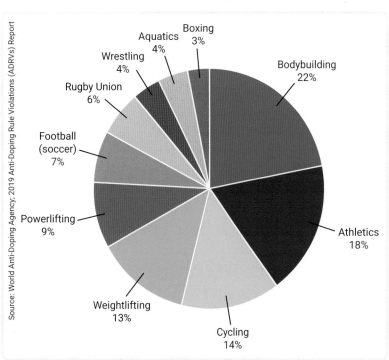

Source: World Anti-Doping Agency; 2019 Anti-Doping Rule Violations (ADRVs) Report

- Bodybuilding 22%
- Athletics 18%
- Cycling 14%
- Weightlifting 13%
- Powerlifting 9%
- Football (soccer) 7%
- Rugby Union 6%
- Wrestling 4%
- Aquatics 4%
- Boxing 3%

▲ **FIGURE 4.7.1** The share of anti-doping rule violations worldwide 2019, by sport

4.7 LEARNING CHECK

1 Chemical testing for banned substances occurs at the Summer Olympic Games. Research the frequency of doping violations at the Olympics for several years. **Create** a bar graph of the number of disqualified athletes who tested positive for a banned substance each year. The x-axis should show Summer Olympic Games by year and the y-axis should show total number of athletes disqualified (men and women combined).

4.8 Using an electronic scale

SCIENCE SKILLS IN FOCUS

IN THIS MODULE, YOU WILL FOCUS ON LEARNING AND IMPROVING THESE SKILLS:

- ▶ use an electronic scale
- ▶ represent data with a graph
- ▶ investigate accuracy of data with percentage error calculations.

In this module, you will investigate how properties of individual metallic elements vary. Density, a physical property, is defined as the ratio of mass to volume. An electronic scale will be used to weigh different metals.

Electronic scales are commonplace in scientific laboratories. They allow for the quick and accurate mass of objects to be obtained.

▲ **FIGURE 4.8.1** Electronic laboratory scales and a weigh dish

▶ **To use an electronic scale:**

1 Ensure the surface of the scale is clean of any debris and/or fingerprints. If needed, gently wipe the surface of the scale clean and dry.

2 Place the scale on a flat and stable surface. The precision of the scale depends on factors such as the stability of the surface the scale rests on.

3 Turn on the scale and wait for the balance screen to display a reading.

4 If you are using a weigh dish to weigh the substance, place the empty weigh dish on the scale and press 'zero' or 'tare' to automatically deduct the mass of the empty weigh dish from the scale.

5 Carefully add the substance to be weighed to the weigh dish. This can be done with the weigh dish on the scale or removed. If removed, be careful not to set the weigh dish in any liquids or grease that will add weight to it. If left on the scale, be careful not to spill any substance on the scale.

6 Place the weigh dish back on the scale and record the mass indicated by the digital display.

Video
Science skills in a minute: Collecting and organising data

Science skills resource
Science skills in practice: Collecting and organising data

INVESTIGATION 1: DENSITY OF COMMON METALS

AIM

To measure the density of several common metals

YOU NEED

☑ cubes of four metallic elements: aluminium, copper, iron, nickel (see Figure 4.8.2)

▲ FIGURE 4.8.2   Cubes of metallic elements

☑ electronic scale
☑ ruler
☑ graph paper
☑ calculator

WHAT TO DO

1   Obtain the mass of each metallic cube by placing it on a zeroed electronic scale. Record your results in a table like the one below.

2   Use the ruler to measure the dimensions (length, width and height) of each cube, recording your results in the table.

3   Calculate the volume of each cube by multiplying length × width × height. Record your answers in the table.

4   Calculate the density of each metal by dividing the mass by the volume. Units are grams per cubic centimetre.

WHAT DO YOU THINK?

1   Look at your table. Which element is the most dense and which is the least dense?

2   Compare your densities with the accepted density values in Table 4.8.1.

▼ TABLE 4.8.1   Accepted density of metals

Metal symbol	Atomic number	Accepted density (g/cm³)
Al	13	2.70
Fe	26	7.84
Ni	28	8.91
Cu	29	8.96

3   If you obtained different densities, propose reasons why your measured value may be different from the accepted value in each case.

Metal element	Mass (g)	Length (cm)	Width (cm)	Height (cm)	Volume (cm³)	Density (g/cm³)

GOING FURTHER: GRAPH YOUR RESULTS AND LOOK FOR A TREND

Within the periodic table, density typically increases down a group. It generally increases across a period as well, but there are some exceptions.

1   Graph your results for the four elements you studied. In science graphs, the independent variable always goes on the *x*-axis, while the dependent variable goes on the *y*-axis. In this experiment, the density is dependent on the type of metal, so density is the dependent variable and belongs on the *y*-axis. The type of metal, or atomic number of the metal, goes along the *x*-axis.

2   In science, we often look for trends in data. A trend is a pattern between variables in the experiment. In this Going further, your variables were density and atomic number. Look at your graph. Do you see any trends or patterns between density and atomic number? Ensure you include both variables in your trend statement.

3   Are there any exceptions to the trend in your data? If so, what are they?

4   In science, we use a calculation called percentage error to see how close our measurements are to the accepted values, or to see how accurate our results are. Calculate the percentage error between your calculated density and the accepted density for each of your metals by using the following equation.

$$\text{percentage error} = \left(\frac{\substack{\text{accepted density} \\ \text{value}} - \substack{\text{your calculated} \\ \text{density value}}}{\text{accepted density value}} \right) \times 100$$

Record your results in a table similar to the one below.

Metal	Your calculated density (g/cm³)	Accepted density (g/cm³)	Percentage error
Al		2.70	
Fe		7.84	
Ni		8.91	
Cu		8.96	

5   What might you conclude from your percentage error values? Are there improvements that could be made in this experiment that might decrease error?

INVESTIGATION 2: 3D MOLECULAR MODELS

AIM

To construct 3D ball-and-stick models

YOU NEED

☑ molecular modelling kit
☑ digital camera (optional)

WHAT TO DO

1   Read the key provided with the modelling kit. Identify the different 'balls' that represent each of the following atoms.

   • carbon
   • nitrogen
   • hydrogen
   • chlorine
   • fluorine
   • sulfur
   • boron

2 Copy the table below.

Name	Structural formula	3D drawing or photo
a hydrogen	H — H	
b hydrogen chloride	H — Cl	
c hydrogen sulfide	S with H and H bonded below	
d boron trichloride	Cl bonded to B, with Cl and Cl below	
e ammonia	H and H bonded to N, with H below	
f trichloromethane	Cl — C — Cl with Cl above and H below	

3 Construct models of each of the substances in the table, and complete the table as you build the models by drawing your model or taking photographs to print later.

WHAT DO YOU THINK?

1 Compare the structural formula with the 3D model identifying any similarities and differences between the two representations.

2 Justify which representation you think is the most helpful to understanding molecules.

REMEMBERING

1 **Recall** the term given to the smallest unit of matter.

2 **Describe** the difference between an atom of an element and a molecule of an element. Use examples to support your answer.

3 **List** the 13 monatomic and diatomic elements.

UNDERSTANDING

4 Copy the cipher below into your notebook. Decode the paragraph that follows by inferring the correct missing word, then transferring the code letter to the cipher. Some letters have been provided. Use your completed cipher to solve the code.

Alphabet letter	A	B	C	D	
Code letter	F				
Alphabet letter	E	F	G	H	
Code letter			W		
Alphabet letter	I	J	K	L	
Code letter					
Alphabet letter	M	N	O	P	
Code letter					
Alphabet letter	Q	R	S	T	U
Code letter			H		
Alphabet letter	V	W	X	Y	Z
Code letter	A	G	M		

PARAGRAPH:
Chemical [OQOXOEUT] are classified on the periodic [UVIQO] according to the similarities in their [MBJMOBUKOT]. They are classified as metals, [EJE-XOUVQT] or metalloids. Metals often have a shiny metallic [QDTUBO] and are good [LJEYDLUJBT]. On the other hand, non-metals are often [YDQQ] and are [EJU] good conductors.

CODE:
WJQY XOQUT VU 1064°L.

5 **Explain** the difference between a molecular compound and an ionic compound.

6 **Recall** how Dmitri Mendeleev organised his first periodic table.

7 **Summarise** the three types of mixtures. Ensure you include a particle diagram for each type.

8 Acetic acid, 20%, is more concentrated than acetic acid, 6%. Draw diagrams to represent these two solutions at the particle level. The solvent in each solution is water.

9 **Describe** how the scientific understanding of elements changed over time.

10 The names of the elements calcium and carbon both begin with the letter 'c.' **Explain** why the chemical symbol for both is not simply 'C'.

APPLYING

11 **Identify** and count the number of atoms in each of the following compounds. Identify each as a molecular compound or an ionic compound. The first row has been completed for you.

	Chemical formula	Number and type of each atom	1st element	2nd element	3rd element	Molecular or ionic compound?
a	NBr_3	N = nitrogen: 1 Br = bromine: 3	Non-metal	Non-metal		Molecular
b	CH_4O					
c	$NaNO_3$					
d	H_2SO_4					

12 Which element is represented on the periodic table by the following locations?

   a Period 2, group 14

   b Period 3, group 16

   c Group 15, period 6

   d Group 10, period 4

ANALYSING

13 The table below shows properties of common elements at 20°C. Use this table to answer the following questions.

   a Determine the chemical symbol for beryllium.

   b **Identify** elements that have a density of more than 3.5 g/cm³.

   c Does germanium have an atomic weight more than or less than that of chromium?

Element	Symbol	Atomic weight (amu)	Density (g/cm³)
aluminium	Al	26.98	2.70
beryllium	Be	9.012	1.85
boron	B	10.81	2.34
bromine	Br	79.90	3.10
cadmium	Cd	112.41	8.65
carbon	C	12.011	2.25
caesium	Cs	132.91	1.87
chromium	Cr	52.00	7.19
cobalt	Co	58.93	8.90
copper	Cu	65.55	8.94
gallium	Ga	69.72	5.90
germanium	Ge	72.59	5.32

CREATING

14 Use the key below to complete the missing sections of the table. Use the correct colours in your answers.

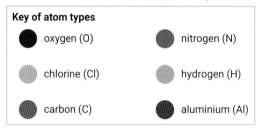

Key of atom types

oxygen (O)   nitrogen (N)

chlorine (Cl)   hydrogen (H)

carbon (C)   aluminium (Al)

Metal	Chemical formula	Structural formula
a		
b	CO_2	
c		
d	$AlCl_3$	
e		
f	NO	

BIG SCIENCE CHALLENGE PROJECT

#4

1 Connect what you have learned

In this chapter you learned about atoms, elements and compounds such as salt, NaCl. Create a mind map to show how the information you have learned is connected.

2 Check your thinking

Explain, in terms of atoms, what sort of substance salt is. If salt is such an iconic substance, can we have too much of it in our diet?

3 Make an action plan

Research how salt is used as a preservative and how this can be detrimental to the human body.

4 Communicate

Use your knowledge and understanding to create an infographic on the amount of salt in our food and health concerns of having too much sodium in our diets.

Getty Images/iStock/Elenathewise

5 Chemical and physical change

5.1 Physical change (p. 118)

A physical change often leads to a change in appearance, not a change in chemical composition.

5.2 Chemical change (p. 122)

A chemical change results in a new substance being formed.

5.3 Evidence of chemical change (p. 125)

Chemical reactions can be identified by several observable factors.

5.4 Energy change in a reaction (p. 128)

Energy in a chemical reaction is stored in the bonds. Heat is a form of energy.

5.5 Properties and uses of substances (p. 131)

Substances can be described by their physical and chemical properties. Properties affect a substance's everyday use.

5.6 FIRST NATIONS SCIENCE CONTEXTS: **First Nations Australians' use of chemical and physical changes** (p. 133)

First Nations Australians developed many sophisticated chemical processes to produce substances that were scarce or unavailable.

5.7 SCIENCE AS A HUMAN ENDEAVOUR: **Biodegradable materials** (p. 136)

Single-use plastics are being replaced with biodegradable products.

5.8 SCIENCE INVESTIGATIONS: **Accurately measuring liquids** (p. 137)

Identifying signs of a chemical change

9780170463027

Getty Images/Stone/Peter Adams

▲ **FIGURE 5.0.1** Tropical rainforests are vital to life on Earth.

Tropical rainforests cover approximately 7 per cent of Earth's surface. They are home to more than half of all the plant species on the planet. Via chemical reactions, these rainforests generate around 20 per cent of the oxygen in Earth's atmosphere and are therefore known as the 'lungs of the planet'. Not only do they generate oxygen, they also remove the heat-trapping gas carbon dioxide (CO_2) from the air.

▶ **How do rainforests produce oxygen and reduce pollutants?**

#5 SCIENCE CHALLENGE ACCEPTED!

At the end of this chapter, you can complete Big Science Challenge Project #5. You can use the information you learn in this chapter to complete the project.

Assessments
- Prior knowledge quiz
- Chapter review questions
- End-of-chapter test
- Portfolio assessment task: Science investigation

Videos
- Science skills in a minute: Measuring liquids **(5.8)**
- Video activities: Changing states of matter **(5.1)**; Energy change of reactions **(5.4)**; Biodegradable plastic **(5.7)**

Science skills resources
- Science skills in practice: Accurately measuring liquids **(5.8)**
- Extra science investigations: Observing physical change **(5.1)**; What do you know about chemical change? **(5.2)**; Chemical change **(5.3)**

Interactive resources
- Label: Phase changes **(5.1)**; Evidence of chemical change **(5.3)**
- Simulation: Phase changes **(5.1)**
- Drag and drop: Chemical or physical change? **(5.3)**; Physical v chemical properties **(5.5)**
- Match: Equations and reactions **(5.2)**

 Nelson MindTap

To access these resources and many more, visit:
cengage.com.au/nelsonmindtap

Video activity
Changing states of matter

Interactive resources
Simulation: Phase changes
Label: Phase changes

Extra science investigation
Observing physical change

physical change
a change in a substance that does not involve the production of a new substance; can usually be reversed

state of matter
one of the forms in which matter can exist: solid, liquid, gas or plasma

GET THINKING

What do mowing the lawn and cracking an egg have in common? They are both examples of physical changes, which are occurring all around you in everyday life. Try to think of some other physical changes happening around you.

A change in appearance, not composition

When a substance undergoes a **physical change**, its appearance changes, but the substance is still the same – no new substance is produced and there is no change in chemical composition of the substance. Physical changes can usually be reversed. You can think of changing a physical property as changing your appearance. You can cut your hair or change your clothes, but you are still the same person as when you started. The change can be reversed by simply growing your hair again or changing back into your former clothes. Let's take a closer look at the most common types of physical changes.

Changes of shape

Changing size or shape is an example of a physical change. If you mould a ball of playdough into a dinosaur shape, it is still playdough, just in a new shape (Figure 5.1.1). You could reverse the process by rolling the dinosaur back into a ball. Changes of shape often occur due to a force being applied; for example, a twist, pull, bend or break.

Changes of state and phase change

A **state of matter** is one of the forms in which matter can exist. There are four natural states of matter: solid, liquid, gas and plasma. For example, water has three states: liquid (water), solid (ice) and gas (water vapour). When in different states, water has different physical properties – for example, water in its liquid form can be poured, but water in its frozen form cannot – but it is still the same substance. (Physical properties will be discussed further in Module 5.5.)

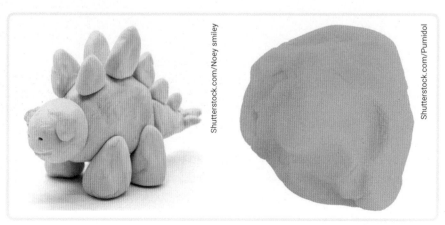

Shutterstock.com/Noey smiley

Shutterstock.com/Pumidol

▲ **FIGURE 5.1.1**   Playdough shaped into a dinosaur is still playdough.

States of matter can be represented by 2D particle models, which are similar to the 2D representations of atoms, elements and compounds discussed in Chapter 4. When a substance changes state, the particles do not change; they just move further apart or closer together. Figure 5.1.2 depicts the three states of water.

Solid
Particles are very close together.

Liquid
Particles are close together.

Gas
Particles are far apart.

▲ FIGURE 5.1.2   The three states of water

A physical change occurs when a substance changes state in the presence or absence of heat. Put very simply, **phase change** refers to this change in a state of matter. When liquid water boils to form gaseous water vapour, its properties change. It is now a gas, and we cannot see it, but it is still water. It could also easily be changed back to liquid water by cooling it.

phase change
a change in the state of matter; an example of a physical change

Common phase changes are shown in Table 5.1.1. These can be represented in a chart, called a phase change diagram, which shows the temperature at which each phase change occurs (see Figure 5.1.3). The *y*-axis is temperature. The horizontal section of the data line indicates regions where there is no temperature change; that is, the temperature remains constant as the change occurs. Freezing and melting points are the same temperature. Condensing and vaporising points are also the same.

▼ TABLE 5.1.1   Types of phase change

Description of phase change	Terminology	Example	
Solid to liquid	**Melting** or liquefying	Ice-cream melting	**melting** changing state from a solid to a liquid
Liquid to solid	**Freezing** or solidifying	Water freezing to ice	**freezing** changing state from a liquid to a solid

▷

▼ TABLE 5.1.1   Types of phase change (continued)

Description of phase change	Terminology	Example
Liquid to gas	**Boiling** or **evaporating** (**vaporising**)	Water boiling in a kettle 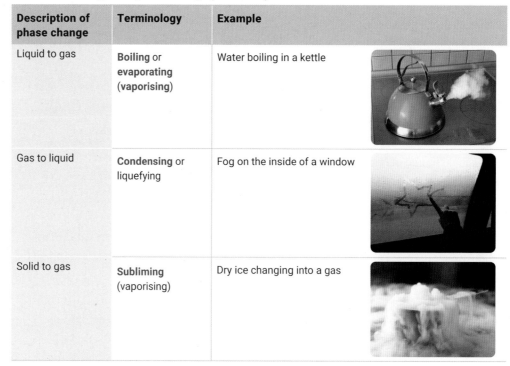
Gas to liquid	**Condensing** or liquefying	Fog on the inside of a window
Solid to gas	**Subliming** (vaporising)	Dry ice changing into a gas

boiling
changing state from a liquid to a gas at a rapid rate

evaporating
changing state from a liquid to a gas

vaporising
changing state from a solid or a liquid to a gas; e.g. evaporating, boiling or subliming

condensing
changing state from a gas to a liquid

subliming
changing state directly from a solid to a gas without going through a liquid phase

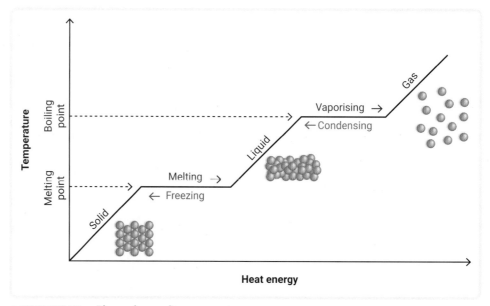

▲ FIGURE 5.1.3   Phase change diagram

Physical change by mixing

A physical change also occurs when two substances are mixed, such as dissolving a solute in a solvent. If solid sugar (solute) is dissolved in water (solvent), we cannot see the sugar but we know it is still there because the resulting solution tastes sweet. If we evaporated the water, we would be left with the original solid sugar. This shows there has been no change to the sugar itself, so it is a physical change. Both liquids and gases can also be dissolved, as shown in Figure 5.1.4.

   9780170463027

▲ **FIGURE 5.1.4** Examples of physical change by mixing: **(a)** sugar dissolving into water; **(b)** dye mixing with water; **(c)** gas bubbles in a carbonated drink

In summary, physical changes:

- involve a change of state, shape or form
- are reversible in most cases
- do not create new substances
- do not alter the chemical properties of a substance, but may alter physical properties, including colour, state or density.

5.1 LEARNING CHECK

1 **List** four examples of physical changes you see around you.
2 **Identify** the two components of a solution.
3 **Recall** the opposite process for each of the following phase changes.
   a Condensing
   b Freezing
4 Figure 5.1.5 shows the particles in a piece of solid chocolate sitting at the bottom of a beaker.

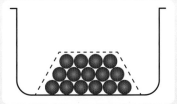

▲ **FIGURE 5.1.5** Particles in a piece of solid chocolate

   a Redraw the diagram to show the arrangement of the chocolate particles after heat has been applied.
   b What evidence could you list to support the idea that a physical change has occurred?
5 Use Figure 5.1.3 (phase change diagram) to answer the following questions.
   a Condensing and what other change in state occur at the same temperature?
   b Which physical state exists at the lowest temperature?
   c Which two processes require the loss of heat from the substance?

5.2 Chemical change

✓ identify and describe the characteristics of a chemical change
✓ explain chemical reactions at a particle level in terms of products and reactants.

Quiz
Chemical change

Interactive resource
Match: Equations and reactions

Extra science investigation
What do you know about chemical change?

> **GET THINKING**
>
> What do a burning log in the fire, a rusting iron nail and baking bread have in common? They are all examples of chemical reactions. Try to think of some other everyday chemical reactions.

A new substance is formed

You have already learned that the chemical composition of a substance is not altered during a physical change. For example, when liquid water is boiled, it changes state from liquid to gas. Liquid water and gaseous water vapour have some different physical properties, such as their state at room temperature, but they are both still made up of identical H_2O molecules. Water vapour can be easily condensed (that is, reversed) back into liquid water by using distillation equipment in the laboratory.

chemical change
when the chemical make-up of a substance changes, and a new substance or substances are formed

chemical reaction
a process that occurs when a substance changes to produce a new substance

reactant
a substance used up in a chemical reaction

product
a new substance produced in a chemical reaction

Unlike physical changes, **chemical changes** are generally not reversible. Chemical change always leads to the production of new/different substances that have different physical and chemical properties from the original substances. A process where a chemical change occurs is called a **chemical reaction**. If a chemical reaction has occurred, the chemical make-up (composition) of the original substance – called the **reactant** – is different from the chemical make-up of the new/difference substance produced – called the **product**.

For example, let's examine what happens when a piece of coal burns. When coal burns, it produces new substances such as soot and gases, including carbon dioxide and water vapour. Coal is not the same substance as soot, ash or carbon dioxide gas, so the new substances (the products) have very different physical and chemical properties from the original piece of coal (the reactant). This means the original substance has reacted and undergone a chemical change. The carbon dioxide, and other substances produced, cannot be turned back into coal again. The coal gets 'used up' and so more coal would need to be burned if we wanted to keep a fire going. (Chemical properties will be discussed further in Module 5.5.) Table 5.2.1 shows other examples of everyday chemical changes.

In summary, chemical changes:

- involve a change in chemical composition
- are often not reversible. If they are reversible, it is by chemical means
- produce new/different substances that have new/different physical and chemical properties.

▼ **TABLE 5.2.1** Everyday examples of chemical changes

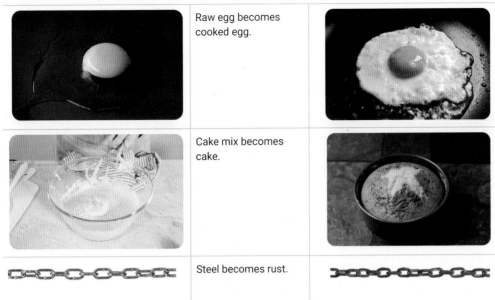

	Raw egg becomes cooked egg.
	Cake mix becomes cake.
	Steel becomes rust.

Representing chemical reactions

Chemical reactions are represented in the same manner by scientists around the world. The reactants are written on the left side of an arrow, while the products are written on the right side of the arrow. The arrow represents the change – that is, the reactants have changed into the products.

$$\text{reactants} \rightarrow \text{products}$$

Chemical reactions can be represented in two different ways. The first method is called a **word equation**, and it contains the full names of the chemicals involved. If there is more than one reactant, a '+' sign is placed between each of the reactants. The '+' sign is similarly used if there is more than one product. For example, in photosynthesis, carbon dioxide and water react to form glucose and oxygen. Carbon dioxide and water are the two reactants, so they are placed on the left side of the arrow with '+' between them. Glucose and oxygen are both products, so they are placed on the right side of the arrow and also have '+' between them. The word equation for this reaction is:

$$\text{carbon dioxide} + \text{water} \rightarrow \text{glucose} + \text{oxygen}$$

word equation
a word summary that shows the reactants and products of a chemical reaction

The second method of representation is called a **chemical equation**. A chemical equation gives the same information as a word equation, except it uses chemical formulas instead of chemical names. The chemical equation for photosynthesis is:

$$6CO_2 + 6H_2O \rightarrow C_6H_{12}O_6 + 6O_2$$

chemical equation
a symbol summary that shows the reactants and products of a chemical reaction

The numbers in front of the chemical symbols are known as coefficients. They show how many atoms or molecules are needed or produced in a chemical reaction. You will learn more about coefficients in Year 9.

Particle models for chemical changes

As described in Module 5.1, particle models are useful to help understand scientific ideas, such as the differences between solids, liquids and gases. Particle models can also be employed to demonstrate chemical changes.

As we learned, when a chemical change occurs, the particles that make up the product are different from those of the reactant. However, it is important to understand what that difference is. Chemical bonds in the reactants are broken, and the atoms rearrange to form bonds in the products. The atoms do not change types or disappear.

Figure 5.2.1 shows how burning magnesium, Mg, reacts with oxygen, O_2, in the air to produce magnesium oxide, MgO. In this reaction, magnesium and oxygen are the reactants and magnesium oxide is the product. The reaction starts and ends with only two types of atoms: magnesium and oxygen. The reactants consist of the magnesium atoms and oxygen molecules as separate substances. After the reaction, the magnesium and oxygen are bonded to each other. This is true for all chemical reactions: the types of atoms don't change, just the arrangement.

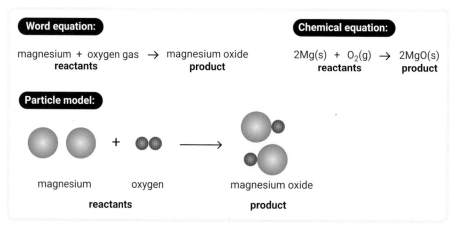

Word equation:

magnesium + oxygen gas → magnesium oxide
reactants **product**

Chemical equation:

$2Mg(s) + O_2(g) → 2MgO(s)$
reactants **product**

Particle model:

magnesium + oxygen ⟶ magnesium oxide
reactants **product**

▲ **FIGURE 5.2.1** Word equation, chemical equation and particle model showing the chemical reaction between magnesium and oxygen

5.2 LEARNING CHECK

1 **List** four examples of a chemical change in your everyday life.
2 **Write** word equations for the following chemical reactions.
   a Carbon reacts with oxygen gas to produce carbon dioxide.
   b Ammonia reacts with hydrochloric acid to produce ammonium chloride.
3 **Compare** chemical changes and physical changes.
4 Figure 5.2.1 shows the chemical reaction for burning a piece of magnesium metal. A photo of this experiment is shown in Figure 5.2.2.
   a **Identify** two safety procedures you would need to undertake if you were to do this experiment.
   b Will the product be different from the reactants in this reaction? **Explain** your reasoning.

▲ **FIGURE 5.2.2**
Burning magnesium

9780170463027

5.3 Evidence of chemical change

BY THE END OF THIS MODULE, YOU WILL BE ABLE TO:
✓ describe the indicators that are evidence of a chemical change
✓ classify change as physical or chemical.

GET THINKING

A burning log fire produces ash and other substances that were not there before. A shiny grey iron nail may turn orange-brown over time as a new substance, called rust, is produced. What other indicators that a chemical change has occurred can you think of?

Interactive resources
Drag and drop: Chemical or physical change?
Label: Evidence of chemical change

Extra science investigation
Chemical change

Indicators of a chemical change

Chemical changes occur at the atomic or molecular level of a substance. Atoms can never be created, and they don't disappear. Rather, the atoms that make up the reactants are rearranged in a different way to form the products, which are new substances that have different chemical properties.

You cannot see atoms or molecules, so how do you know if a chemical change has occurred? When combining two substances, sometimes nothing happens or something happens so slowly that it appears as though nothing is happening. Other times, something will happen immediately to show that a chemical change has occurred (Figure 5.3.1).

Have you ever done the pop test to verify the presence of hydrogen gas? Instructions for this experiment are included in this chapter's Science Investigations. If you collect hydrogen gas in a cold test tube and place a lit taper over the opening of the test tube, you should hear a 'pop'. This shows the presence of flammable hydrogen gas. It is an example of a chemical change. If you look closely at the inside of the top of the test tube, you may see some colourless drops of liquid. This is water vapour that has condensed on the cold test tube. It shows that when you combine hydrogen gas (H_2) with oxygen gas (O_2), a new substance, water (H_2O), is produced.

Chemical reactions can often be identified by several observable factors. Table 5.3.1 lists some indicators that a chemical change has occurred.

Shutterstock.com/iamlukyeee

▲ **FIGURE 5.3.1** The combustion of fireworks is an example of an immediate chemical change because heat, light and sound are produced when the firework fuse comes into contact with the charcoal-based black powder that acts as a fuel source.

▼ **TABLE 5.3.1** Some indicators that a chemical change has occurred

Reactant atoms rearrange to form:	You will observe:
a product that is a different colour.	a permanent colour change.
a substance that is a gas.	gas (bubbles) given off and/or there may be an odour.
a new solid substance.	an **insoluble** solid, called a **precipitate**.
a new substance with different energy.	a temperature change, which can be hotter or cooler, and/or light is given off.

insoluble
not able to be dissolved

precipitate
a solid substance formed in a solution as a result of a chemical reaction

A colour change

Chemicals often have a characteristic colour. When a new substance is formed during a chemical reaction, it may be a different colour from the reactants. If you leave a half-eaten apple on the table, it will turn brown as it begins to react with oxygen in the air. A shiny grey iron nail may turn orange-brown over time as a new substance, called rust, is produced. These are examples of a colour change during a chemical reaction. However, a colour change does not always indicate that a chemical change has occurred. For example, adding red food colouring to blue creates purple colouring, but this is only a physical change.

Gas is produced

When reactant atoms rearrange to form a product that is a gas, several things can occur. Sometimes the product has a smell. This is a sign that a chemical change has occurred. A common example of this is the smell rotten food gives off. This smell is quite different from the smell given off when the food was fresh.

Gases can form bubbles in a chemical reaction. An antacid tablet releases bubbles when dropped into a glass of water. When the antacid tablet is in solid form, the two chemicals, commonly citric acid and bicarbonate, cannot react with each other. However, when they are dissolved in water, a chemical reaction between the two occurs and carbon dioxide gas, CO_2, is given off.

An insoluble solid: precipitate

A precipitate, shown in Figure 5.3.2, is an insoluble substance that can form when two solutions chemically react. The solid substance (the precipitate) makes that liquid appear cloudy. Over time, the solid often settles to the bottom.

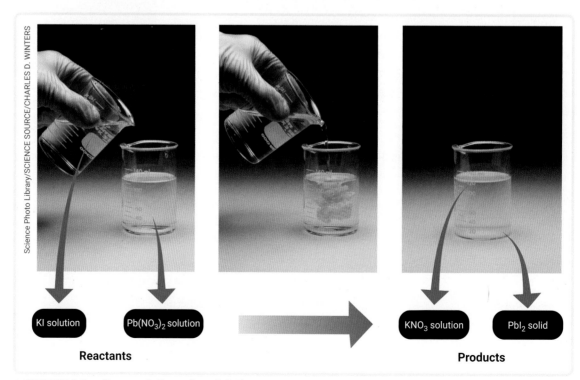

Science Photo Library/SCIENCE SOURCE/CHARLES D. WINTERS

KI solution   Pb(NO₃)₂ solution   KNO₃ solution   PbI₂ solid

Reactants   **Products**

▲ **FIGURE 5.3.2**   Representations of precipitation

Temperature change

As described earlier, during a chemical reaction, existing bonds are broken and new bonds form. When this occurs, energy is released or absorbed in the form of heat. A common example of this is fire. The heat produced by fire can be used to cook food. We will go into more detail on this topic in the Module 5.4.

5.3 LEARNING CHECK

1 **List** the indicators that signify a chemical change has occurred.

2 **Classify** each of the diagrams below as either a physical or a chemical change.

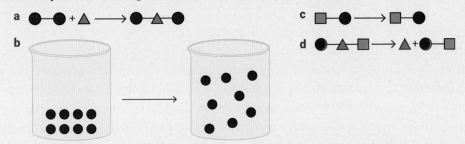

3 Two students each added a level spatula of purple crystals into 100 mL of water in an evaporating basin. The water turned purple and was transparent. Next, they heated the liquid until no more liquid was left. A purple powder (solid) was left in the evaporating basin. One student said that when they added the purple crystals to the water there was a chemical reaction, while the other said there was not.

   a **Describe** the reasons why the first student thought it was a chemical reaction.

   b **Describe** the reasons why the second student thought it was not a chemical reaction.

   c Which student is correct? **Explain** your answer.

4 **Analyse** each photo and identify the evidence that indicates a chemical change has occurred.

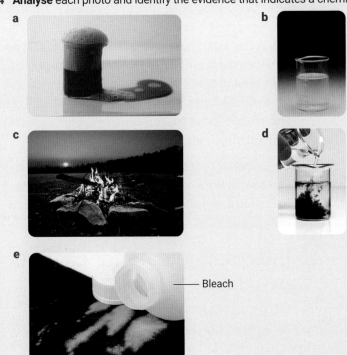

Bleach

5.4 Energy change in a reaction

BY THE END OF THIS MODULE, YOU WILL BE ABLE TO:

✓ analyse data to provide evidence that a chemical change has occurred

✓ explain how differences in temperature indicate that a chemical change has occurred.

Video activity
Energy change of reactions

GET THINKING

Have you ever wondered why a cold pack in a first aid kit becomes cold so quickly, after simply snapping it? That is an example of an endothermic reaction working for us!

Energy in a chemical reaction

In Chapter 4 you learned that atoms can chemically bond to other atoms. We expanded that knowledge in Module 5.2, discussing how bonds between some (but not all) atoms will break and re-form as atoms rearrange during a chemical reaction. However, before the atoms of the reactants can rearrange themselves to form new products, the bonds between the atoms in the reactants must be broken, and it takes energy to pull them apart. Put simply, chemical bonds store energy. Stored energy is called potential energy. Energy stored in the chemical bonds of a substance is called **chemical potential energy**. Chemical potential energy:

chemical potential energy
energy stored in the chemical bonds of a substance and released when the substance reacts

- allows animals, such as humans, to obtain energy from the chemical reactions that occur in digesting food
- is released by batteries
- produces energy when fossil fuels are burned.

Heat is a form of energy

endothermic reaction
a chemical reaction that takes in heat energy

exothermic reaction
a chemical reaction that releases heat energy

Chemical reactions involve either absorbing or releasing energy. Chemical reactions that absorb energy in the form of heat are called **endothermic reactions**. These reactions take in heat from the surroundings, and that heat is used as the energy for the reaction to occur. In this type of reaction, the temperature will drop. A common example is the instant cool pack found in first aid kits (Figure 5.4.1). When the pack is squeezed, the inner bag of water is broken, and the water dissolves the solid in the pack. The process absorbs heat from the surroundings, which lowers the temperature of the pack.

Chemical reactions that release energy are called **exothermic reactions**. These reactions give off energy in the form of heat. Their temperature increases. A common example is burning wood.

Interpreting data for change in heat

Consider a classroom experiment using two different chemicals: chemical A and chemical B. Chemical A

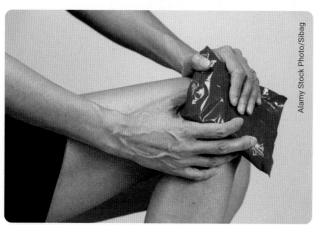
Alamy Stock Photo/Sibag

▲ **FIGURE 5.4.1** An instant cool pack is an example of an endothermic reaction

combines with water in an endothermic reaction and chemical B combines with water in an exothermic reaction.

The experiment

Each chemical was mixed in a different beaker with the same volume of water. A thermometer was used to measure the temperature of the water before and after the chemical and water were mixed. The data was graphed as shown in Figure 5.4.2. Can you tell which line represents chemical A? Which line represents chemical B?

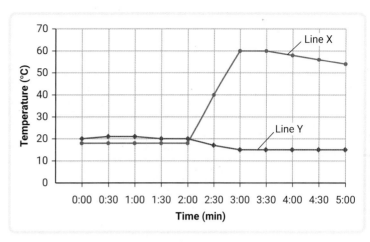

▲ **FIGURE 5.4.2** Temperature change during an experiment using two different chemicals

Analysing the data

Both solutions remained at a constant temperature, around 20°C, for the first 2 minutes. Then, the temperature of one of the solutions (line X) rose from 20°C to 60°C over 1 minute while the temperature of the other solution (line Y) dropped from 20°C to 15°C in the same amount of time. Line X showed an increase in temperature, meaning it was an exothermic reaction. Therefore, chemical B is represented by line X. Line Y showed a decrease in temperature, meaning it was an endothermic reaction. Therefore, line Y represents chemical A.

5.4 LEARNING CHECK

1 **Recall** whether an endothermic reaction or an exothermic reaction releases heat energy.
2 **Explain** how, during a chemical reaction, a change in energy could be measured.
3 Figure 5.4.3 is a graph showing temperature change during a chemical reaction. Is the reaction endothermic or exothermic? **Explain** your reasoning.

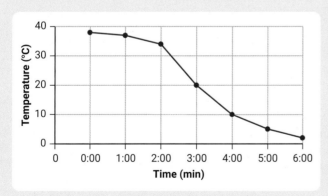

▲ **FIGURE 5.4.3** Temperature change during a chemical reaction

4 Analyse the graph in Figure 5.4.4 to answer the following questions.

   a At what time does reaction A reach 40°C?

   b What temperature is the mixture in reaction B at 3 minutes?

   c What is the initial temperature for both reactions?

   d Classify each reaction as endothermic or exothermic. Justify your answer.

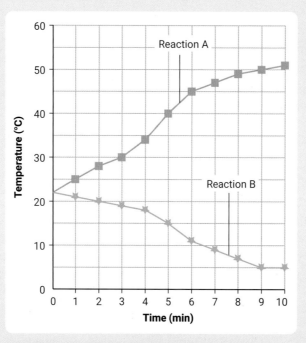

▲ FIGURE 5.4.4   Temperature change in two chemical reactions

5 Analyse the exothermic reaction data in the table below and do the following.

   a Fill in the missing data with approximate values.

   b Calculate the temperature change for the reaction.

Time (min)	Temperature (°C)
0	22
2	
	26
6	28

   c Construct a graph using the data from your completed table. The *x*-axis should be time (independent variable); the *y*-axis (dependent variable) should be temperature.

6 Recall three ways chemical potential energy plays a role in everyday life.

5.5 Properties and uses of substances

BY THE END OF THIS MODULE, YOU WILL BE ABLE TO: ✓ explain the relationship between a substance's properties and its use.

Interactive resource
Drag and drop: Physical v chemical properties

Substances can be described by their properties

In a cutlery drawer, there are separate compartments to separate the knives, spoons and forks. Utensils are classified in this way according to what they are used for. Materials can also be classified according to the properties they have in common. Scientists divide properties into two categories: physical and chemical.

Chemical properties

The features of a substance that determine how it reacts with other substances are called **chemical properties**. You cannot observe the chemical properties of a substance the way you can the physical properties. But if two substances have similar chemical properties, you will be able to tell by observing how they both react with the same substance. If they react in a similar way, they have similar chemical properties.

chemical properties
the properties of a substance that determine how it reacts when combined with other substances

Chemical properties include:

- **flammability** (the ability to catch fire)
- **toxicity** (how poisonous a substance is)
- the ability to **corrode**
- sensitivity to light (which, for example, causes newspaper to turn yellow over time).

flammability
the ability of a substance to catch fire

toxicity
a measure of how poisonous a substance is

corrode
the breakdown of a metallic substance due to a chemical reaction with chemicals in the environment, such as oxygen or water

Consider this example. Paper will easily catch fire but a substance called asbestos does not. However, paper is unlikely to harm you, whereas asbestos is highly toxic and causes diseases such as lung cancer when the fibres have been inhaled. Paper and asbestos have different chemical properties. Paper is flammable and non-toxic, whereas asbestos is not flammable but is toxic.

Physical properties

Physical properties are features of a substance that can be observed or examined without changing the composition of a substance. Examples are colour, hardness, density, and melting and boiling points. You have already investigated many of these in previous modules.

physical properties
the properties of a substance that can be observed or examined without changing its composition

▲ FIGURE 5.5.1
The properties of aluminium make it well-suited to span long distances between power poles.

Properties affect the use of a substance

The properties of a substance can determine what it is used for. For example, a raincoat is made of waterproof material because its purpose is to keep you dry in the rain. You would not use a plastic pan to fry bacon because it would melt. Instead, you would use a metal pan because it transfers heat quickly and can withstand high temperatures.

Substances used in business and industry are also chosen for their properties. For example, a paper coffee cup with ridges on its outer layer can hold very hot liquids because paper is a good insulator.

Copper wiring is used in electrical wiring in homes, but it is not commonly used to span long distances between power poles. Instead, a form of aluminium, like that used for a soft drink can, is used. Aluminium has properties that are better suited for this purpose. It conducts electricity almost as well as copper, but it is half the density and costs less to manufacture (Figure 5.5.1).

☆ ACTIVITY

Useful properties of common substances: think–pair–share

What to do

1 Copy the table below.

Item	Chemical or physical property of the item that makes it suitable for its purpose
Glass food jar	Transparent, non-toxic
Plastic table tennis ball	
Garden hose	
Stainless steel in building construction	
Plastic rainwater tank	

2 **Identify** the chemical and/or physical properties of each item that makes it suitable for its purpose. The first has been done for you.

3 Add more rows to your table, with everyday objects around your home.

What do you think?

1 Pair with another student or two and discuss what you wrote down.

2 Share your results with the class.

5.5 LEARNING CHECK

1 **List** three chemical properties.

2 Think of something you use to get ready for school. How are its properties suited for its use?

3 Use the data in Table 5.5.1 to **compare** the density of gaseous and liquefied natural gas.

▼ TABLE 5.5.1 Properties of liquefied natural gas (LNG)

Natural gas chemical formula	CH_4
Boiling point	−161°C
Liquid density	426 kg/m³
Gas density at 25°C	0.656 kg/m³

9780170463027

5.6 First Nations Australians' use of chemical and physical changes

IN THIS MODULE, YOU WILL:

✓ explore First Nations Australians' use of physical and chemical changes.

Resins

Before colonisation, First Nations Australians observed and experimented with natural substances. This experimentation led them to develop many sophisticated chemical processes, which allowed them to produce required substances that were either scarce or unavailable.

Resins are produced by plants. Resins were, and still are, widely used by First Nations Australians as adhesives for manufacturing and repairing tools. The type of resin depends on the plant source and its availability. Heating makes the resin soft and malleable. This allows it to be easily applied to the surfaces that are to be cemented together. Once the resin has cooled, the surfaces are firmly bound together. If a tool requires repair, the resin holding the parts together can be reheated and reshaped. The parts are put back into position and the resin is left to cool and harden again. If too much heat is applied to the resin, it will become brittle when it cools and cannot be resoftened. The resin would burn if it becomes too hot.

Resins and gums from plants could also be burned to produce fire for burning Country, or to provide light at night. Records show that on Yidinjdji country in the Cairns–Yarrabah area of northern Queensland, a form of slow-burning torch was made by coating a branch with resin from the black kauri pine and setting it alight.

Fermentation

First Nations Australians developed fermentation processes using native yeasts and bacteria, sugar-rich saps and nectars to produce alcoholic drinks. For example, cider gum, a Tasmanian eucalypt, provided a sugar-rich sap used by the Palawa People to produce a fermented drink, while the Noongar People from south-western Western Australia used nectar from certain banksia species to produce their drink.

Pyrolysis

Pyrolysis is the process of heating plant material to very high temperature, without oxygen. The plant material gets so hot that

▲ **FIGURE 5.6.1** Many First Nations Australians have sustainably harvested grasstree resin over thousands of years.

▲ **FIGURE 5.6.2** Scientists from the University of Adelaide and Australian Wine Research Institute collect sap from the Tasmanian cider gum (*Eucalyptus gunnii*). This sap has traditionally been used to produce a fermented drink called *wayalinah*.

it breaks down. First Nations Australians used pyrolysis to produce numerous important substances.

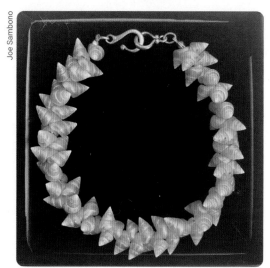

▲ FIGURE 5.6.3 A *marina* shell bracelet made by Palawa Elder Jeanette James Lutrawita (Tasmania)

- Some First Nations Australian groups in central Australia developed a medicine that helped people when travelling long, difficult journeys. They found that the addition of quicklime (calcium oxide) increased the effectiveness of the medicine. First Nations Australians used pyrolysis of plants high in calcium oxide to produce the required quicklime ash.
- Charcoal was produced for use in pigments and added to adhesives.
- Pyroligneous acid was made by some Palawa women of Tasmania and used to dissolve the thin outer coating on *marina* shells. This exposed the highly iridescent layer underneath. The shells were then used to make necklaces and other jewellery. (Figure 5.6.3).
- Wood ash was produced from plants high in sodium and potassium and was used as a substitute for sea salt (sodium) by First Nations Australian groups living far from the sea. They understood that sodium was vital for many physiological processes.

☆ ACTIVITY 1

1 Draw up a table with the headings given below. Add several blank rows below the heading row. Use the information provided in this module to identify at least four different processes involving physical and/or chemical changes.

Starting substance (reactant)	Product(s)	Process	Chemical or physical change?	Reason

▲ FIGURE 5.6.4 First Nations Australians used controlled heating (calcination) to change gypsum into plaster.

Plaster

Historical records show First Nations Australians producing plaster for toys, games and body ornamentation. They used a process of controlled heating (called calcination) to change the naturally occurring gypsum (calcium sulfate) into plaster. First Nations Australians understood the importance of carefully controlling the amount of heat in the production of plaster. They knew that if too much heat was used, the final product would not set after water was added. Records also show the production of pigments for paints using the same method.

9780170463027

In this investigation you will compare the physical and chemical properties of gypsum before and after calcination to determine whether a physical or chemical change has occurred.

 Warning

Wear safety glasses and protective equipment. Do not touch hot equipment with bare hands. Follow your teacher's instructions about disposal of waste material.

You need
- 15 g gypsum
- electronic balance
- mortar and pestle
- crucible and tongs
- pipe clay triangle
- Bunsen burner and tripod
- 6 test tubes
- test-tube rack
- spatula
- distilled water
- 10 mL 0.5 mol/L hydrochloric acid
- matches

What to do

1 Place 15 g of gypsum in a mortar and pestle and grind it to a powder.

2 Record the physical properties of the powder produced.

3 Place a small spatula of the powder in each of three test tubes in a test-tube rack and label the test tubes 'Gypsum 1', 'Gypsum 2' and 'Gypsum 3'. Ensure there is enough powder left for the next step.

4 Place the remaining powder in a crucible, and then place the crucible on a pipe clay triangle on a tripod over a Bunsen burner.

5 Light the Bunsen burner, turn it to a blue flame and heat the powder for 2–3 minutes.

6 Turn off the Bunsen burner and allow the crucible and product to cool.

7 Record the physical properties of the product.

8 Place a small spatula of the product in each of three test tubes in a test-tube rack and label the test tubes 'Product 1', 'Product 2' and 'Product 3'.

9 One-third fill each test tube labelled '2' with distilled water, and shake it gently to mix. Feel the test tubes. Record your observations.

10 One-third fill each test tube labelled '3' with hydrochloric acid, and shake it gently to mix. Feel the test tubes. Record your observations.

11 Compare the appearance of the substances in the test tubes labelled '1'. Record your observations.

What do you think?

1 Review all your observations and determine whether the calcination of gypsum results in a physical or a chemical change. Support your decision with evidence from your observations.

2 **Suggest** how you could determine whether the gypsum had been overheated in the investigation, producing an unusable product.

5.7 Biodegradable materials

BY THE END OF THIS MODULE, YOU WILL BE ABLE TO:

✓ explain how science communication has helped the community form views on the use of biodegradable materials.

Video activity
Biodegradable plastic

science communication
talking, raising awareness or even arguing about science-related topics

biodegradable
describes a substance or product that is able to completely break down and return to natural products within a short time

Some takeaway restaurants have replaced plastic straws with cardboard straws, and plastic forks, knives and spoons with wooden eating utensils. Some shopping outlets have eliminated single-use plastic bags. Do you know why?

Single-use plastics are being replaced because of their harmful environmental impact. They don't readily break down and they contaminate land and the oceans. You may have read about this topic or had discussions about it with others. Talking, raising awareness, or even arguing about science-related topics is called **science communication**.

In 2021, the Australian Government released a plan for industry to phase out the use of single-use non-recyclable plastic. This includes plastics that, if littered, have a negative environmental impact. The government published the policy, raised awareness and advocated for the policy. These were forms of science communication.

Single-use plastics have been replaced with similar products made of **biodegradable** materials. Biodegradable means the product completely breaks down and returns to natural products within a short time – typically a year or less. Items such as cardboard straws and wooden utensils break down or degrade naturally, molecule by molecule. The smaller pieces can assimilate back into the environment because they have the same composition as the natural material that they were made from; for example, timber.

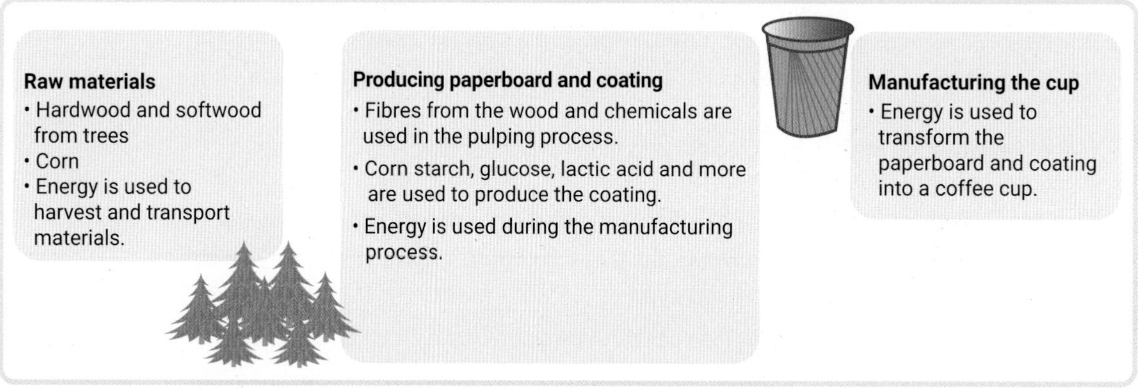

Raw materials
- Hardwood and softwood from trees
- Corn
- Energy is used to harvest and transport materials.

Producing paperboard and coating
- Fibres from the wood and chemicals are used in the pulping process.
- Corn starch, glucose, lactic acid and more are used to produce the coating.
- Energy is used during the manufacturing process.

Manufacturing the cup
- Energy is used to transform the paperboard and coating into a coffee cup.

▲ **FIGURE 5.7.1** What natural products go into making a biodegradable coffee cup?

▲ **FIGURE 5.6.4**
Bioplastic is a biodegradable plastic.

5.7 LEARNING CHECK

1 Biodegradable plastics also exist. Figure 5.7.2 shows an example of this. **Research** the following claim: 'Bioplastic cups are better for the environment than plastic cups.' Ensure your research uses data.

2 Think and make notes about your viewpoints on biodegradable materials and discuss them with your classmates.

SCIENCE SKILLS IN FOCUS

IN THIS MODULE, YOU WILL FOCUS ON LEARNING AND IMPROVING THESE SKILLS:

▶ accurately record the volume in a measuring cylinder by reading the meniscus.

Liquids can be difficult to measure, even with scientific glassware such as a measuring cylinder. In this Science skills in focus, you will learn what a **meniscus** is and how you can accurately measure liquids in a measuring cylinder by correctly reading a meniscus.

A meniscus is the downward or upward curve at the top of a liquid close to the surface of the container, as shown below in Figure 5.8.1.

Alamy Stock Photo/Science Photo Library

a   b

▲ FIGURE 5.8.1   (a) Mercury forms an upward curve at the meniscus; (b) water forms a downward curve at the meniscus.

The meniscus is caused by **surface tension** between the liquid molecules and the container. To get an accurate reading, you need to read the volume at eye level with the meniscus, as shown in Figure 5.8.2.

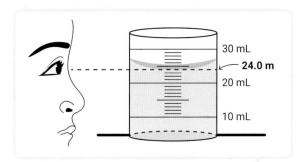

30 mL
24.0 m
20 mL
10 mL

▲ FIGURE 5.8.2   Read the volume at eye level with the meniscus to get an accurate reading.

▶ To get an accurate reading:

- Set the measuring cylinder on a stable surface and bring your eyes level with the height of the liquid.
- Take the measurement at the centre of the meniscus; that is, at its lowest or highest point, as shown by the dashed line in Figure 5.8.2.
- When the liquid falls between two graduations (lines on the measuring cylinder), you estimate that digit.
- In Figure 5.8.2, the reading would be 24.0 mL.

meniscus
the downward or upward curve at the top of a liquid caused by surface tension and adhesion between the liquid molecules and the container

surface tension
the tension of the surface of a liquid, caused by the attraction of the particles in the surface layer by the bulk of the liquid, which tends to minimise surface area

Video
Science skills in a minute: Measuring liquids

Interactive resource
Science skills in practice: Accurately measuring liquids ·

IDENTIFYING SIGNS OF A CHEMICAL CHANGE

AIM

To practise reading a meniscus in a graduated cylinder and record and analyse observations from various experiments to determine chemical change

Working in a small group, conduct each of the experiments below and identify any signs of a chemical reaction. Record your observations in a table like the example shown.

Experiment	Meniscus readings (mL)	Gas produced	Temperature reading (°C)		Precipitate	Colour change	Other
			Initial	Final			
1							
2							
3							
4							

The volume needed for each experiment varies and does not need to be exact. However, aim to get close to the approximate volume and record your meniscus reading in the table, ensuring that you estimate if the meniscus falls between graduation lines.

TIPS

☑ Label each test tube with the experiment number before you begin.

☑ Ensure your measuring cylinder is rinsed or cleaned before each experiment.

FOR ALL EXPERIMENTS, YOU NEED

☑ test-tube rack
☑ 5 × 25 mL Pyrex glass test tubes
☑ markers to label the test tubes
☑ thermometer

Other materials are listed within each experiment.

Experiment 1

YOU NEED

☑ 25 mL measuring cylinder
☑ 10 mL of 2 mol/L hydrochloric acid
☑ magnesium ribbon
☑ match and taper (for alternative option)

Warning

Treat all chemicals as toxic and avoid contact with eyes and skin. Wear safety glasses and follow any other precautions or instructions given to you by your teacher.

WHAT TO DO

1 Use the measuring cylinder to measure out approximately 10 mL of hydrochloric acid into test tube 1 and place it in the test-tube rack.

2 Place the thermometer into the test tube and record the initial temperature of the hydrochloric acid.

3 Add a curled up 5 cm strip of magnesium ribbon to the test tube. Swirl the test tube gently and replace the thermometer.

4 Record the highest or lowest temperature reached after three minutes. Feel the test tube. Does it feel warmer or cooler?

5 Record all observations.

As an alternative option, conduct a hydrogen pop test.

1 Follow steps 1 and 3 above, skipping step 2. Place a chilled, empty test tube over the top of test tube 1, to collect the hydrogen gas produced.

9780170463027

2 Collect gas for around 1 minute, then carefully remove the top test tube, keeping it inverted.

3 Pass a lit taper under the opening of the removed test tube. Record all observations. Water is evidence that a new substance was produced.

Experiment 2

YOU NEED

- ☑ 10 mL measuring cylinder
- ☑ sodium bicarbonate
- ☑ 5 mL vinegar
- ☑ spatula

WHAT TO DO

1 Use the clean measuring cylinder to measure 5 mL of vinegar. Pour it into test tube 2 and place it in the test-tube rack.

2 Place the thermometer into the test tube and record the initial temperature of the vinegar.

3 Use a spatula to add about half a scoop of sodium bicarbonate to the test tube.

4 Record the highest or lowest temperature reached. Feel the test tube. Does it feel warmer or cooler?

5 Record all observations.

Experiment 3

YOU NEED

- ☑ 10 mL measuring cylinder
- ☑ 5 mL of 0.1 mol/L solution of silver nitrate, $AgNO_3$
- ☑ 5 mL of 0.1 mol/L solution of potassium chloride, KCl

WHAT TO DO

1 Using the measuring cylinder, add 5 mL of silver nitrate solution to test tube 3 and place it in the test-tube rack.

2 Place the thermometer into the test tube and record the initial temperature.

3 Using a clean measuring cylinder, add 5 mL of potassium chloride into the same test tube.

4 Record the highest or lowest temperature reached. Feel the test tube. Does it feel warmer or cooler?

5 Record all observations.

Experiment 4

YOU NEED

- ☑ 10 mL measuring cylinder
- ☑ 5 mL of 0.5 mol/L sodium hydroxide, NaOH
- ☑ 5 mL of 0.5 mol/L hydrochloric acid, HCl
- ☑ universal indicator solution in a dropper bottle

WHAT TO DO

1 Use a clean measuring cylinder to add 5 mL of 0.5 mol/L sodium hydroxide to test tube 4 and place it in the test-tube rack.

2 Place the thermometer into the test tube and record the initial temperature.

3 Add two drops of universal indicator solution to the test tube.

4 Record your observations.

5 Use a clean measuring cylinder to add 5 mL of 0.5 mol/L hydrochloric acid to test tube 5.

6 Place the thermometer into test tube 5 and record the initial temperature.

7 Add two drops of universal indicator to test tube 5.

8 Record your observations.

9 Pour the contents of test tube 4 into test tube 5.

10 Record the highest or lowest temperature reached. Feel the test tube. Does it feel warmer or cooler?

11 Record all observations.

WHAT DO YOU THINK?

1 Analyse your data from each experiment to determine whether a chemical reaction occurred. Be sure to justify your choice by explaining your reasoning with data.

CONCLUSION

Write a conclusion that summarises your findings. Make sure it relates to your aim.

5 REVIEW

REMEMBERING

1 **Recall** where the energy that is released during a chemical reaction is stored.

2 **Recall** whether the products or reactants are on the left side of the arrow in a chemical equation.

3 **Define** 'precipitate'.

UNDERSTANDING

4 Imagine that you could see the atoms moving around as a substance changed appearance. **Describe** how you would know whether the process is a physical or a chemical change.

5 Imagine tearing a piece of paper in half, putting the halves on top of each other, and then tearing them again. If you repeated this again and again, the paper pieces would get smaller but the stacks would get thicker. **Explain** whether this is a physical or chemical change.

6 **Classify** each of the characteristics as representative of a chemical or physical change.

   a It is easily reversed.

   b The components can be separated by physical means; for example, sifting.

   c It involves breaking bonds and forming new bonds between atoms.

   d New chemical substances are formed.

7 **Explain** why it is necessary to know the properties of a material before we use it. Give an example.

8 **Discuss** what is happening in the diagram below.

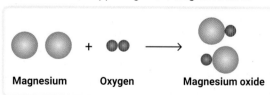

Magnesium   Oxygen   Magnesium oxide

9 Using a particle diagram, **explain** why ice melting to become liquid water is a physical change.

10 **Explain** why a difference in temperature can indicate that a chemical reaction has occurred.

APPLYING

11 The following observations were made during a class experiment.

   • The solid is powdery white and has a tangy smell.

   • It ignites and burns when heated.

   a **List** the physical properties of the powder.

   b **List** the chemical properties of the powder.

12 Chemical A was combined with Chemical B. The observations are recorded below. For each of the three examples, **explain** whether you think a chemical change has occurred, and why.

	Chemical A	Chemical B	Products
a	Clear solution	Clear solution	Cloudy and milky
b	Clear solution	Clear solution	Clear solution
c	Clear solution	Pale orange solution	'Blood red' solution

13 Two unknown solutions are combined and observed. The data is shown below. **Identify** whether a chemical reaction occurred and explain why or why not.

	Chemical A	Chemical B	Product
Colour	Red	Blue	Two layers: one red and one blue

ANALYSING

14 Simon mixed two chemicals in a flask and placed the flask on an electronic balance that had already been zeroed for the mass of the flask. Simon measured the combined mass of the substances every minute for 7 minutes. He observed bubbles on the surface of the combined substances throughout the experiment. The table below shows the data he collected.

Time (minutes)	Mass of two substances (g)
0	10.0
1	9.9
2	9.7
3	9.6
4	9.5
5	9.5
6	9.4
7	9.3

a Use the data in the table to **construct** a simple line graph.

b **Identify** the trend in the data by describing what is happening to the mass as time increases.

c Calculate the change in mass over 7 minutes.

d Use your graph to **estimate** the mass at 3.5 minutes.

e Use your graph to **estimate** the mass of substance that would remain after 8 minutes.

f **Justify** whether this experiment represents a physical or a chemical change.

15 The table below shows the observed results of four experiments. The experimenter marked a cross when she observed each listed item. The absence of a cross means that phenomenon was not observed. Determine which experiment(s) resulted in chemical change.

Experiment	Bubbles and/or smell	Colour change	Temperature change	Precipitate
1	X		X	
2				X
3				
4		X		

16 Lithium is sometimes called white gold due to its high value. It has a range of uses in both chemical and technical applications due to its physical and chemical properties. The major technical application for lithium is in the production of ceramics and glasses. This includes heat-resistant glass and ceramics such as those used in ovenwear and cooktops.

a Use the pie chart below to determine the following.

i What percentage of lithium production worldwide is used in ceramics and glass?

ii What is the second largest use for lithium?

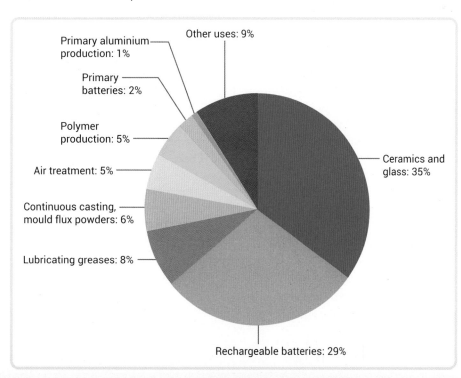

Primary aluminium production: 1%
Other uses: 9%
Primary batteries: 2%
Polymer production: 5%
Air treatment: 5%
Continuous casting, mould flux powders: 6%
Lubricating greases: 8%
Ceramics and glass: 35%
Rechargeable batteries: 29%

9780170463027

b The table below shows Australia's lithium production and reserve life. Use this to answer the following questions. Note: kt Li = kilotonnes of lithium.

Year	Ore reserves (kt Li)	Production (kt Li)
2017	1662	21.3
2016	1361	14
2015	851	n/a
2014	854	n/a
2013	854	n/a
2012	854	12.7
2011	506	11.7
2010	174	7.3
2009	58	5.8

© Commonwealth of Australia (Geoscience Australia) 2021 (CC BY 4.0)

i In what year did Australia produce (mine) the most lithium?

ii Is there a trend in the ore reserves from 2009 to 2017?

17 First Nations Australians used the processes of pyrolysis and calcination to produce important substances. **Compare** these two processes, **identifying** the physical and/or chemical changes involved.

EVALUATING

18 The recycling of many materials is based on the physical properties of the material. During the recycling process for aluminium, which can occur an infinite number of times, the aluminium goes through several physical changes. The stages are listed below, out of order. Determine the correct order and re-write the steps.

A Chips are melted and re-formed into solid blocks of aluminium.

B Bales are cut and smashed into small aluminium chips.

C Blocks are pressed and rolled into thin aluminium sheets.

D New cans are cut and moulded.

E Empty, used cans are collected by people.

F Used cans are crushed and formed into large cubic bales.

19 The symbol for recycling is shown below. **Evaluate** how this symbol is representative of recycling.

Shutterstock.com/CalypsoArt

CREATING

20 Construct a flow chart using the following terms: solid, liquid, gas, melting, freezing, condensing, evaporating, subliming.

BIG SCIENCE CHALLENGE PROJECT

#5

1 Connect what you have learned

In this chapter, you have learned about chemical reactions and indicators that chemical changes have occurred. Create a mind map to show how the information you have learned is connected.

2 Check your thinking

a Explain, in terms of chemical change, what happens during photosynthesis.

b Examine the reactants and products and think about how this chemical reaction decreases the pollutant carbon dioxide, CO_2, while producing oxygen.

c What evidence is present during photosynthesis to indicate that a chemical change is occurring?

d Use your answers to deduce why tropical rainforests are called the 'lungs of the planet'.

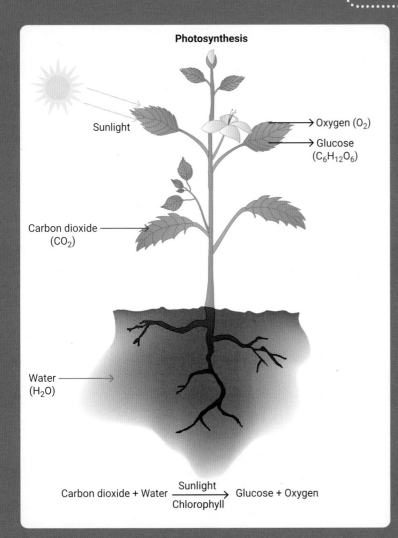

Photosynthesis

Sunlight

Oxygen (O_2)

Glucose ($C_6H_{12}O_6$)

Carbon dioxide (CO_2)

Water (H_2O)

$$\text{Carbon dioxide} + \text{Water} \xrightarrow[\text{Chlorophyll}]{\text{Sunlight}} \text{Glucose} + \text{Oxygen}$$

3 Make an action plan

Conduct research into the claim that photosynthesis is one of the most important chemical reactions on Earth.

4 Communicate

Use your knowledge and understanding to create a persuasive text addressing the above claim. Your text should contain data from a reliable source.

6 The rock cycle

6.1 Rocks and minerals (p. 146)

Rocks and minerals are the building blocks of Earth's surface.

6.2 Igneous rocks (p. 150)

Igneous rocks are formed from liquid rock.

6.3 Weathering and erosion (p. 154)

Weathering and erosion change Earth's surface.

6.4 Sedimentary rocks (p. 158)

Sedimentary rocks are made from the products of weathering and erosion.

6.5 Metamorphic rocks (p. 160)

Metamorphic rocks form when other rocks are changed by heat and pressure.

6.6 The rock cycle (p. 164)

The rock cycle describes the formation of and changes to rocks, minerals, crystals and other materials.

6.7 Fossils (p. 166)

Fossils are the remains and traces of living things preserved in rock.

6.8 FIRST NATIONS SCIENCE CONTEXTS: **First Nations Australians' shaping of rocks into tools** (p. 170)

First Nations Australians have a deep understanding of the properties of rocks and minerals, which they applied to the manufacture of stone artefacts.

6.9 SCIENCE AS A HUMAN ENDEAVOUR: **Dealing with acid mine water** (p. 173)

Mine rehabilitation is essential to reducing the impact that mining has on the environment.

6.10 SCIENCE INVESTIGATIONS: **Fieldwork** (p. 174)

1 Planning to conduct fieldwork on Country/Place

2 Mineral research

▲ FIGURE 6.0.1 Volcanic bumpy boulder on Mars

This image was taken on Mars by the NASA Spirit rover. It shows a boulder thrown out of a Martian volcano. Do you think the rocks on Mars are very different from those on Earth? Is it possible that life once lived on Mars? If so, how can we know?

In the future, humans may journey to Mars to study the planet and perhaps live there. How can our understanding of the rocks on Earth help us to understand the nature and history of Mars?

▶ **Are you up for the challenge of learning about the geology of Mars?**

 #6 **SCIENCE CHALLENGE ACCEPTED!**

At the end of this chapter, you can complete Big Science Challenge Project #6. You can use the information you learn in this chapter to complete the project.

Assessments:
- Prior knowledge quiz
- Chapter review questions
- End-of-chapter test
- Portfolio assessment task: Project

Videos
- Science skills in a minute: Representing data **(6.10)**
- Video activities: Weathering **(6.1)**; Rock cycle **(6.6)**; How do fossils form? **(6.7)**; Cultural reconnection: Mine rehabilitation **(6.9)**

Science skills resources
- Science skills in practice: Representations of data **(6.10)**
- Extra science investigations: Crystal formation **(6.1)**; Modelling the rock cycle **(6.6)**

Interactive resources
- Drag and drop: Igneous rock classification **(6.2)**; How sedimentary rocks form **(6.4)**; Which rock is which? **(6.5)**
- Label: Volcanic structures **(6.2)**; The rock cycle **(6.6)**; Fossil formation steps **(6.7)**
- Crossword: Rock types **(6.5)**

❄️ Nelson MindTap

To access these resources and many more, visit:
cengage.com.au/nelsonmindtap

BY THE END OF THIS MODULE, YOU WILL BE ABLE TO:
✓ identify minerals based on their properties
✓ explain the relationship between rocks, minerals and ores, and give examples of each.

GET THINKING

What do you know about rocks and minerals? Write two questions you would like to learn the answers to from this module.

Video activity
Weathering

Extra science investigation
Crystal formation

rock
a naturally occurring solid made up of minerals

Rocks and minerals

We see **rocks** every day, but we rarely see them change. Although rocks are not living, they do change over long periods of time and can change from one type to another (Figure 6.1.1). In this chapter, you will learn about rock types as well as the forces and factors that cause them to change. But first, let's focus on answering the question: what is a rock?

Rocks are naturally occurring solids made up of substances called **minerals**. A mineral is a naturally occurring **inorganic** solid. There are thousands of types of minerals, and some of them are valuable as gems (Figure 6.1.2).

Sandstone is a common rock (Figure 6.1.3). It contains grains of the mineral quartz and is held together by other minerals (Figure 6.1.4). In sandstone, minerals such as quartz and calcite act as a cement, binding the rock together.

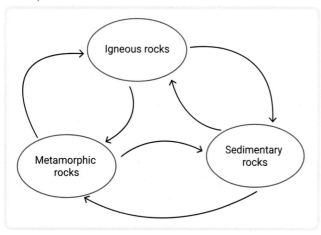

▲ **FIGURE 6.1.1** Rocks can change from one type to another over long periods of time.

mineral
a naturally occurring inorganic solid with a neatly ordered crystal structure and characteristic composition

inorganic
a substance not formed from the remains or products of living things

amethyst · analcime · anapaite · apatite · aquamarine · celestine
chabazite · danburite · fluorite · galena · halite · heliodor
hematite · kyanite · morganite · orpiment · petalite · phlogopite
pyrite · realgar · saamit · tanzanite · thomsonite · uvarovite

Shutterstock.com/vvoe

▲ **FIGURE 6.1.2** A variety of minerals

▲ FIGURE 6.1.3   A sample of sandstone

▲ FIGURE 6.1.4   Sandstone as seen under a microscope. The actual width of the section shown is about 2.25 mm.

Mineral structure

Minerals are defined by an ordered arrangement of atoms and a characteristic chemical composition. A mineral can consist of a single element or a chemical compound. Gold, sulfur and carbon (diamond and graphite) are naturally occurring elements that are considered to be minerals. But most minerals are chemical compounds and, therefore, can be represented by a chemical formula. For example, fluorite (Figure 6.1.5) has the formula CaF_2, meaning that it consists of calcium (Ca) and fluorine (F) with twice as many fluorine atoms as calcium atoms.

A mineral's orderly internal structure is called a **crystal** structure. Atoms, or groups of atoms, are arranged in a three-dimensional repeating pattern called a crystal lattice. Figure 6.1.6 shows how calcium and fluorine atoms are arranged in a crystal lattice in fluorite. Mercury, a liquid metal, is the only naturally occurring non-crystalline mineral.

The crystal structure of a mineral determines its physical properties. Sometimes minerals can have the same chemical composition, but, because they have atoms arranged in different ways, they have different physical properties and are therefore considered to be different minerals. For example, graphite and diamond are both minerals made of carbon. In fact, most things on Earth, including people, contain carbon! But when you arrange

▲ FIGURE 6.1.5 The mineral fluorite

crystal
a solid in which the atoms are arranged in a well-ordered pattern

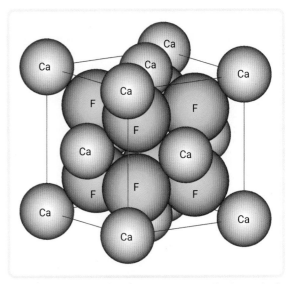

▲ FIGURE 6.1.6   The regular arrangement of calcium (Ca) and fluorine (F) atoms in the mineral fluorite

Chapter 6 | The rock cycle

Graphite
Soft, dull, opaque, common

Diamond
Hard, shiny, transparent, rare

▲ **FIGURE 6.1.7** (a) Graphite and (b) diamond are minerals with the same chemical composition but different crystal structures.

atoms in different ways you can get different structures. Graphite, the black material in a pencil, is quite soft because the atoms are arranged loosely. In a diamond, however, all the carbon atoms are connected strongly to each other in a three-dimensional structure, making the mineral very hard (Figure 6.1.7).

How minerals are identified

A mineral is identified by its physical properties. Some physical properties are described in Table 6.1.1.

▼ **TABLE 6.1.1** Properties of minerals

Property	Description
Colour	Colour can be a useful guide to some minerals. Colour can be produced by small amounts of other elements added to a crystal structure.
Streak	Streak is the colour of a mineral when it is turned into a fine powder. Streak is produced with softer minerals by scraping them across a tile.
Hardness	Hardness refers to how easy it is to scratch a mineral. **Mohs scale of hardness** (named after the scientist who created the scale) consists of 10 standard minerals (Figure 6.1.8). If one of the 10 standard minerals scratches the unknown mineral, the standard mineral is harder. If it does not, the unknown mineral is the same hardness or harder.
Cleavage	Many minerals tend to split, or cleave, along flat surfaces. These surfaces are called cleavage. There can be more than one direction of cleavage in a mineral.
Lustre	Lustre refers to how light interacts and reflects from the surface of a crystal, rock or mineral. Some minerals are dull, but minerals such as quartz reflect light like glass, and are said to have a glassy lustre.
Density	Density is how tightly packed the atoms are in something. If something is dense, the atoms are very close together. If something is not dense, the atoms are loosely packed.

Mohs scale of hardness
a scale used to measure the relative hardness and resistance to scratching between minerals

cleavage
the way a mineral splits to produce a flat surface

lustre
how light looks when reflected from the surface of a crystal, rock or mineral

9780170463027

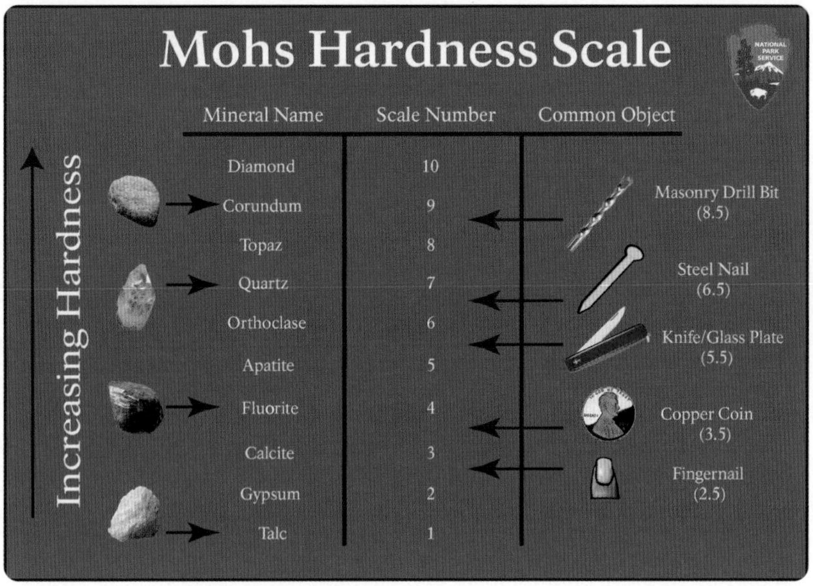

Mohs Hardness Scale

Mineral Name	Scale Number	Common Object
Diamond	10	
Corundum	9	Masonry Drill Bit (8.5)
Topaz	8	
Quartz	7	Steel Nail (6.5)
Orthoclase	6	Knife/Glass Plate (5.5)
Apatite	5	
Fluorite	4	Copper Coin (3.5)
Calcite	3	
Gypsum	2	Fingernail (2.5)
Talc	1	

Increasing Hardness

National Park Service/U.S. Department of the Interior

▲ **FIGURE 6.1.8** Mohs scale of hardness

Ores, minerals and their uses

Some rocks contain one or more valuable minerals that can be extracted, processed and then sold at a profit (Figure 6.1.9). Valuable minerals are referred to as **ore**, or the ore minerals. Usually rocks are a mixture of many different minerals, so the valuable ore has to be separated from the less valuable minerals, called **gangue minerals**.

Minerals are important for several reasons. They are a resource from which we obtain metals and other useful substances. For example, we use unaltered minerals such as sapphires and rubies as gemstones. Elements from minerals provide essential nutrients for plants and animals. Minerals are also important because they tell us about the origin and history of rocks, which helps us understand Earth and Earth's history.

Getty Images/E+/BeyondImages

▲ **FIGURE 6.1.9** Iron ore is a valuable resource mined in Western Australia. This open cut mine is in the remote Pilbara region.

ore
useful minerals that can be extracted and processed to be sold

gangue minerals
minerals found in rocks that are less valuable than ore

6.1 LEARNING CHECK

1 **Define** 'mineral'.
2 **Describe** what defines a mineral.
3 **Identify** the minerals found in the rock called sandstone.
4 **Describe** four physical properties that are used to identify minerals.
5 A mineral is found to scratch a sample of calcite but cannot scratch a piece of glass. What is the hardness of this mineral on Mohs hardness scale?
6 **Describe** the relationship between rocks and minerals.
7 What is the difference between an ore mineral and a gangue mineral?
8 **Explain** how you would tell whether you had a sample of a rock or a mineral.

Interactive resources
Drag and drop: Igneous
rock classification
Label: Volcanic
structures

magma
extremely hot liquid or
semi-liquid rock formed
under the surface of Earth

lava
hot, molten rock that is
expelled during a volcanic
eruption

volcano
an opening in Earth's
crust, through which
molten rock reaches the
surface

igneous rock
a type of rock formed
when molten materials
(magma or lava) cool and
solidify

GET THINKING

What do you think of when you read or hear the word lava? How many types of material can liquid rock become when it cools?

The formation of igneous rocks

Magma is hot liquid or semi-liquid rock formed under the surface of Earth. Magma rises from deep within Earth through the crust. If magma reaches Earth's surface, it loses its dissolved gases and is called **lava** (Figure 6.2.1). Lava flows from a volcanic eruption. (You will learn more about **volcanoes** in Chapter 7.) When magma or lava cools, it crystallises and solidifies to form solid rock. This type of rock is known as **igneous rock**.

▲ **FIGURE 6.2.1** A lava flow showing a rapidly cooled surface broken by the fluid lava underneath

The mineral crystals in igneous rocks interlock with each other. The size of the crystals in an igneous rock depends on how quickly the liquid rock has cooled. Magma may cool slowly, kilometres below the surface of Earth. As a result, the crystals formed are large – big enough to see without a microscope. When lava cools on the surface of Earth, it cools very quickly and the crystals are very small – too small to see with your eye alone.

Intrusive igneous rocks

Igneous rocks formed from magma deep under the surface of Earth are called **intrusive** rocks. It may take thousands or even millions of years for the magma to become solid, because the earth around the magma keeps it warm. Examples of intrusive rocks are granite, diorite, dolerite and gabbro. Some of the spectacular cliffs along the south-eastern coast of Tasmania are composed of dolerite (Figure 6.2.2).

Extrusive igneous rocks

Extrusive rocks crystallise on Earth's surface. If lava erupts in water, a thin, solid skin will form in seconds. This may trap gas bubbles, which create spherical holes in the resulting rock. On land, the lava surface may solidify in minutes, but the inside of the lava flow takes longer to cool. In Hawaii, the average lava flow thickness is 10–15 metres and it can take up to a year and a half to fully cool and solidify. Examples of extrusive rocks are basalt, andesite, **obsidian** and rhyolite. Obsidian is an extrusive rock that cools very rapidly (Figure 6.2.3). It is a volcanic glass because it cools so fast that crystals do not have a chance to form.

Figure 6.2.4 shows some of the structures formed from magma and lava. **Batholiths** are very large volumes of intrusive rock formed from solidification of magma. They form at least 2 kilometres below Earth's surface. Volcanic ash forms when cooling lava is shattered by escaping gas, forming tiny fragments of glass. Ash and larger rock fragments may be joined together, and the resulting rocks are a type of extrusive rock called **pyroclastic** rock.

intrusive
describes an igneous rock formed from magma below the surface of Earth

extrusive
describes an igneous rock formed from lava at or above the surface of Earth

obsidian
a volcanic glass

batholith
a very large volume of intrusive rock, formed deep under Earth's surface by solidification of magma

pyroclastic
describes rock formed from volcanic ash and rock fragments

◀ **FIGURE 6.2.2** Dolerite cliffs such as these on Tasmania's Cape Hauy are the tallest sea cliffs in the southern hemisphere – up to 300 metres tall.

▼ **FIGURE 6.2.3** Obsidian – a glass formed by extremely rapid cooling of lava

Chapter 6 | The rock cycle   **151**

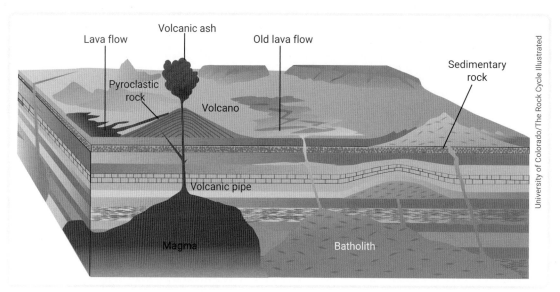

▲ FIGURE 6.2.4 Igneous and volcanic structures

Identifying igneous rocks

Igneous rocks are identified by their interlocking crystals. They are classified according to the minerals they contain (composition) and their grain size (texture). Some magmas produce minerals rich in silicon, aluminium and potassium. The rocks that form are light in colour (Figure 6.2.5), often containing glassy quartz and a pink mineral called feldspar. Other magmas produce rocks that are rich in minerals containing iron, magnesium and calcium. Such rocks tend to be dark (Figure 6.2.6). Rocks, such as granite, with crystals large enough to see, are called coarse grained. Rocks, such as basalt, with crystals too small to see are called fine-grained.

▲ FIGURE 6.2.5 Granite – an intrusive rock. The individual mineral crystals are 3–4 mm in diameter.

▲ FIGURE 6.2.6 Basalt – an extrusive rock

9780170463027

▼ TABLE 6.2.1 Igneous rocks are classified by composition and texture

Composition	Composition name	Light coloured	Intermediate	Dark coloured
	Minerals present	Quartz, potassium and sodium feldspar, mica	Feldspar, hornblende, biotite	Pyroxene, calcium-rich feldspar, some olivine
Texture (grain size)	Coarse-grained examples	Granite	Diorite	Gabbro
	Fine-grained examples	Rhyolite	Andesite	Basalt
	Porphyry (mixed crystal sizes)	• Named after the large crystals present; for example, feldspar-porphyry • Can be named for the type of rock; for example, a porphyritic basalt		
	Glassy	• Obsidian – a solid volcanic glass • Pumice – a rock formed from gas trapped in glass		
	Pyroclastic (fragments)	• Volcanic breccia – angular fragments larger than several millimetres across • Tuff – rock formed from ash fragments less than 2 mm in diameter		

6.2 LEARNING CHECK

1 Look at the rocks shown in Figures 6.2.5 and 6.2.6. What does the difference in crystal size indicate about where the rocks formed?

2 What do all igneous rocks have in common?

3 **Explain** why rocks formed in the batholith contain larger crystals than the rocks in a lava flow.

4 Locate a building or monument in your area made of igneous rock. **Describe** the minerals in the rock and the size of the mineral crystals.

5 The rocks in Figure 6.2.7 are granites that once formed kilometres beneath the surface. How did they reach the surface?

▲ FIGURE 6.2.7

6.3 Weathering and erosion

BY THE END OF THIS MODULE, YOU WILL BE ABLE TO:

✓ describe how weathering and erosion change rocks and create sediments.

weathering
the breakdown of minerals and rocks at Earth's surface by physical or chemical processes

GET THINKING

Look at the key words in the margins of this module. What do you already know about these terms/processes? Can you identify examples where these processes have affected landscapes in your area?

▲ **FIGURE 6.3.1** This rock is being weathered. How can you tell?

Weathering

Rocks do not exist forever. Zircon minerals dated as being more than 4 billion years old have been found embedded in rocks at Jack Hills in Western Australia. However, these rocks are not as old as the minerals inside them. So where are the original rocks that housed the minerals?

The breaking down of rocks and minerals by natural processes is called **weathering**. Scientists believe that the minerals found at Jack Hills originated in rocks that have weathered (and metamorphosed – see Module 6.5) over time.

☆ **ACTIVITY 1**

Introduction to weathering

You need

- course-grained granite
- magnifying glass
- mortar and pestle

> ⚠ **Warning**
> Always wear safety glasses during this experiment and listen to your teacher's instructions. Remember to grind the rock; don't pound it.

What to do

1 Observe a sample of coarse-grained granite with a magnifying glass. How many different minerals can you see? Describe them in your notes.

2 Grind a small sample of granite to a coarse sand using a mortar and pestle.

3 Observe the ground-up granite with a magnifying glass. How have the minerals changed?

4 Describe your observations in your notes.

5 Keep the ground-up granite for activity 2.

What do you think?

1 Did all the minerals grind up in the same way? Can you suggest a reason why?

2 Why is this an example of weathering?

9780170463027

Physical weathering

The effects of changing temperature on rocks, or the effects of water, can break a rock into smaller pieces. This is called **physical weathering**. Physical weathering is common in cold conditions or where temperatures change a lot each day.

A common type of physical weathering is freeze–thaw, or ice wedging. Water seeps into cracks in a rock during the day and freezes into ice at night. The ice expands as it forms, eventually splitting the rock (Figure 6.3.2).

Salt crystals growing between mineral grains can also cause rock to break down. Wind may carry sand or ice, which wears down and polishes rocks. Rocks may also undergo physical weathering when other rocks are removed from above them. Removing the top rocks leads to a release in pressure, causing the underlying rocks to expand and split.

Chemical weathering

When the chemical composition of minerals in a rock are changed by air or water, this is called **chemical weathering**. Chemical weathering occurs when minerals in the rock are changed by chemical reactions. Water can dissolve minerals. Acids can change minerals too. The red and yellow colours of some rocks are due to iron reacting with oxygen (Figure 6.3.1).

physical weathering
a process of weathering that breaks rocks apart or wears them down, but does not change their chemical composition

chemical weathering
a process of weathering that changes the chemical composition of the minerals in rocks

Alamy Stock Photo/Nature Picture Library

▲ **FIGURE 6.3.2** Freeze–thaw weathering occurs as ice wedges open cracks in a rock.

Investigating chemical weathering

☆ **ACTIVITY 2**

You need

- ground granite from activity 1
- mortar and pestle
- water
- wash bottle
- filter funnel
- filter paper
- medium-sized test tube
- test-tube rack
- 10 mL measuring cylinder

Warning

Always wear safety glasses during this experiment and listen to your teacher's instructions. Remember to grind the rock; don't pound it.

What to do

1 Add about 5 mL of water to the ground granite in a mortar and grind the material for 3 minutes. Record what you see in your notes.

2 Place the filter paper in the filter funnel. Place the funnel in the top of a medium-sized test tube that is sitting in the test-tube rack. Wash all the sediment into the filter paper with a wash bottle.

3 Observe the liquid in the test tube and the material on the filter paper. Describe what you see in your notes.

What do you think?

1 The white precipitate is the result of water reacting with feldspar. Why is this a chemical reaction?

2 **Discuss** with your teacher how you might test for the presence of salts in the water once the precipitate settles.

▲ FIGURE 6.3.3 Biological weathering. How will the tree roots split the rock?

biological weathering
a process of weathering in which living organisms break down rocks

erosion
the movement of weathered material away from where it forms by water, wind, ice or gravity

sediment
solid material transported from one place and deposited in another

transportation
the movement of solid particles by agents of erosion over large distances

deposition
the laying down of sediment

Biological weathering

Living things can also break down rocks. This is called **biological weathering**. Biological weathering takes many forms. Some are physical processes and others are chemical processes. Plant roots can split rocks (Figure 6.3.3). Lichens can release acids that dissolve some minerals. Bacteria and fungi can also chemically weather minerals.

Erosion

Weathering usually produces three types of material: rock and mineral fragments, clay minerals and dissolved salts. **Erosion** is what happens when the weathered materials move away from their original location. Wind, water, ice and gravity can relocate weathered material. These are called agents of erosion. Wind can carry fine sand and dust; ice can move rocks the size of a house. Erosion usually happens slowly, over many years, but sometimes in it can happen suddenly (Figure 6.3.4).

Transportation

The solid material being transported in erosion is called **sediment**. The movement of sediment by agents of erosion over large distances is called **transportation**. Water can carry sediment in rivers or move material by wave action on a coast. Over time, rock fragments on a steep slope can slide downhill. As rocks are transported, they change. For example, as rocks are carried down a river, they become rounder, smaller and more spherical due to abrasion by other rocks and mineral grains (Figure 6.3.5). Soft minerals (for example, mica or calcite) are ground down by harder rocks until only hard minerals such as quartz remain.

Deposition

Deposition is the process in which sediments, soil and rocks build up at a location. In fast flowing water, heavy sediments such as pebbles can settle out. Slow flowing water allows lighter sediment such as sand grains to settle out. Very still water allows tiny mud and clay particles to settle to the bottom. Environments of sediment deposition are shown in Figure 6.3.6.

▲ FIGURE 6.3.4 Rock falls can happen very quickly.

▲ FIGURE 6.3.5 What has rounded these pebbles?

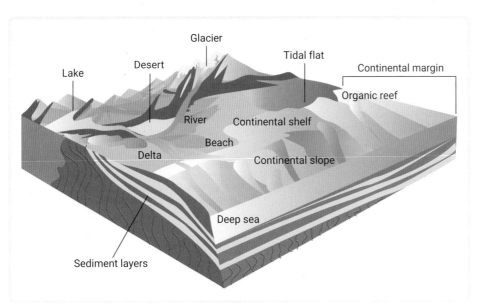

▲ **FIGURE 6.3.6** Common sedimentary environments

Sediment settling

 ✩ **ACTIVITY 3**

You need

- 250 mL measuring cylinder
- dry soil
- water
- plastic wrap

What to do

1 Fill the 250 mL measuring cylinder to the 40 mL mark with dry soil.

2 Fill the measuring cylinder to the 70 mL mark with water.

3 Cover the top of the measuring cylinder with plastic wrap.

4 Hold the base of the measuring cylinder with one hand and cover the top with your other hand. Shake the measuring cylinder vigorously.

5 Place the measuring cylinder on a bench and observe what happens.

> **! Warning**
> Always wear safety glasses during this experiment and listen to your teacher's instructions.

What do you think?

1 What did you observe?

2 Why do you think this happened?

3 What made the water coloured? Will it stay this way?

4 Summarise what the results of the activity show about deposition of sediments.

6.3 LEARNING CHECK

1 **Describe** the process of weathering using an example.

2 **List** the agents of erosion.

3 **Explain** the relationship between weathering and erosion.

4 **Compare** weathering and erosion in terms of the agents involved.

5 **Explain** how the speed of flowing water is related to the types of sediment moved by the water.

Quiz
Types of weathering

6.4 Sedimentary rocks

BY THE END OF THIS MODULE, YOU WILL BE ABLE TO:
✓ describe the origins and types of sedimentary rocks

Interactive resource
Drag and drop: How sedimentary rocks form

GET THINKING

Skim the headings and images of the two pages of this module. What do the headings and images tell you about this module?

How sedimentary rock forms

Sedimentary rock is formed from the products of weathering and erosion. Small pieces of rocks and minerals, clays and salts can all form sedimentary rocks. The remains of once-living organisms can also be buried in sediments and become rock.

Sediments such as sand or mud become rock when they are compacted and cemented together. Sediment is laid down in water in layers. Over time, more sediment is deposited on top. As the sediment builds up, the weight of the sediment forces particles closer together. This process is called **compaction**. Although the sediment grains are close together, water can still move through tiny spaces between the grains. Minerals dissolved in the water crystallise on the surfaces of the sediment, forming cement that binds the sediment together. This process is called **cementation** (Figure 6.4.1). Because sedimentary rocks form from layers of sediment deposited on top of one another, sedimentary rocks form horizontal layers called **beds** (Figure 6.4.2).

sedimentary rock
rock formed from sediments

compaction
the process whereby pressure forces particles closer together

cementation
the process whereby new minerals bind sediment grains together

bed
a horizontal layer of sedimentary rock

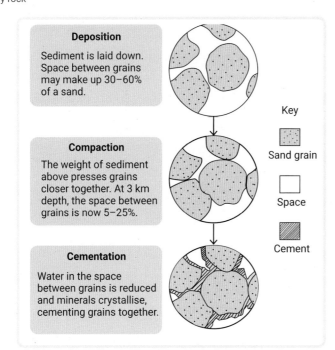

Deposition
Sediment is laid down. Space between grains may make up 30–60% of a sand.

Compaction
The weight of sediment above presses grains closer together. At 3 km depth, the space between grains is now 5–25%.

Cementation
Water in the space between grains is reduced and minerals crystallise, cementing grains together.

Key

Sand grain

Space

Cement

▲ FIGURE 6.4.1 How sedimentary rocks form

▲ **FIGURE 6.4.2** Can you see bedding in the sandstone cliffs? The tree-covered slopes below the cliff are made up of softer sandstones, mudstones and coal layers.

Shutterstock.com/Leah-Anne Thompson

158 Nelson Science 8 | Australian Curriculum

9780170463027

Types of sedimentary rock

Sedimentary rocks can be grouped according to the type of sediment they contain.

- **Clastic** sedimentary rocks: These are made of rock or mineral fragments (clasts) cemented together (Figure 6.4.3). Sandstone is a clastic rock formed from sand-sized particles. Mudstone is formed from clay minerals. Conglomerate is a clastic rock containing rounded pebbles. A layered mudstone is called shale.

- Chemical sedimentary rocks: These are rocks formed from minerals precipitated from water. Some limestones and rock salt are formed in this way.

- **Organic** sedimentary rocks. These rocks are composed of the remains of once-living organisms. Coal is formed from the remains of plants buried in swamps or lakes (Figure 6.4.4). Many microscopic plants as well as coral animals build shells and hard parts, which can form limestones (Figure 6.4.5). Chalk is a limestone formed from the skeletons of microscopic plants and animals.

clastic
small pieces of rock or mineral

organic
relating to, or made from, living material

Sometimes a sedimentary rock can contain a combination of particles, remains of living things and chemicals. For example, oil shale contains clay minerals but also oil, which has formed from the remains of living things. Clastic rocks often contain fossils – the remains of once-living organisms. We will learn more about fossils in Module 6.7.

▲ **FIGURE 6.4.3** Conglomerate contains rounded pebbles.

▲ **FIGURE 6.4.4** Coal is a rock formed from buried plant material and other sediment.

▲ **FIGURE 6.4.5** Limestone is a hard sedimentary rock made of the mineral calcite. It is often formed from the remains of living things. Note the shell impression in this sample.

6.4 LEARNING CHECK

1 **Define** 'sedimentary rock'.
2 Draw a flow chart to show the processes/steps by which a sedimentary rock forms.
3 **Compare** the composition of a sandstone and a coal.
4 **Distinguish** between the three groups of sedimentary rocks.
5 **Explain** why sedimentary rocks might contain the remains of once-living organisms while igneous rocks do not.

Interactive resources
Drag and drop:
Which rock is which?
Crossword: Rock types

GET THINKING

How are metamorphic rocks different from other rocks? Scan the module and make a list of differences you notice.

Types of metamorphism

The minerals in a rock will change over long periods if the environment of the rock changes. Rock exposed at Earth's surface changes through weathering. When rock is buried deep, the heat and pressure conditions there also cause the rock to change. **Metamorphic** rocks are rocks that have been altered by heat, pressure or hot fluids (Figure 6.5.1).

When rocks are altered just by heat, this is called thermal metamorphism or **contact metamorphism**. While rocks buried deep in the earth can experience high temperatures, most contact metamorphism is due to heat from magma. When magma rises in the crust, its heat alters the rocks surrounding it. Minerals recrystallise, changing their size and shape. Soft sedimentary shale becomes a hard, tough, dark metamorphic rock called hornfels (Figure 6.5.2). Sandstone, a sedimentary rock, recrystallises into a harder rock called quartzite (Figure 6.5.3). Soft limestones become hard, sturdy marble.

metamorphic
describes rock that has been altered by heat, pressure or hot fluids

contact metamorphism
rock metamorphism caused by heat

Shutterstock.com/vvoe

▲ **FIGURE 6.5.1** Eclogite – a metamorphic rock with a composition similar to volcanic basalt. Eclogites form when ocean crust is buried deep below the surface of Earth.

▲ **FIGURE 6.5.2** Hornfels – a contact metamorphic rock formed from a clay-rich rock such as shale

▲ **FIGURE 6.5.3** Quartzite – a dense, hard and glassy rock formed from sandstone

A narrow sheet of magma forced through a crack may only change a centimetre of rock next to it. But near a huge batholith, rocks may be changed for hundreds of metres around the igneous rock.

Hot fluids also change rocks. This is known as **hydrothermal metamorphism**. Sometimes hot fluids from large magma bodies travel through nearby rocks and chemically change them. The hot fluids contain lots of dissolved minerals and other elements. The fluids change the chemical composition of the rocks they pass through and cause new minerals to form.

When heat *and* pressure produce metamorphic rocks, this is known as **regional metamorphism**. Where mountains form, rocks experience both heat and enormous pressure. Pressure is the force acting on a surface. We will find out the source of such pressure in Chapter 7. The types of rocks formed depend on the amount of heat and pressure, the original rock, and how long it takes the change to happen.

hydrothermal metamorphism
rock metamorphism caused by hot fluids in the earth changing the chemical composition of rocks they pass through

regional metamorphism
rock metamorphism caused by heat and pressure

Features of metamorphic rocks

Two characteristics of regional metamorphic rocks are layering and the growth of new large crystals in the rock. The layering is called **foliation**. Clay minerals, like the ones found in a mudstone rock, regrow into minerals known as mica minerals. They grow at right angles to the direction of where the pressure is coming from. Minerals do this because it is easier to grow in that direction (Figure 6.5.4). An example of a rock that does this is a schist (Figure 6.5.5). Minerals growing at right angles means the rock forms a layered structure that can cause it to break in sheets.

foliation
layering in a rock formed by crystal regrowth

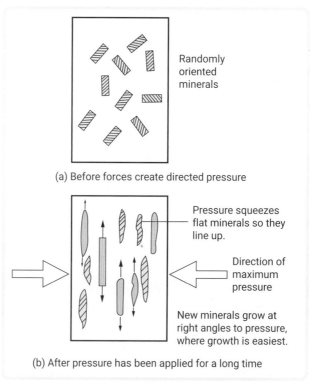

(a) Before forces create directed pressure

Pressure squeezes flat minerals so they line up.

Direction of maximum pressure

New minerals grow at right angles to pressure, where growth is easiest.

(b) After pressure has been applied for a long time

▲ **FIGURE 6.5.4** New minerals grow at right angles to the direction of pressure on them.

▲ **FIGURE 6.5.5** Schist is a rock showing layering.

▲ **FIGURE 6.5.6** Gneiss – this rock forms under intense pressure and at high temperatures.

9780170463027

Type of metamorphism	Original rock type	New rock type	Rock properties
Contact	Limestone	Marble	• Hard • Dense • Mineral crystals have regrown larger
	Sandstone	Quartzite	
	Mudstone	Hornfels	
Hydrothermal	Basalt	Greenstone	• New minerals present • Some hydrothermal rocks contain valuable ore minerals
Regional	Mudstone	Slate	• Foliated • Increasing mineral size • New metamorphic mineral grows in the rocks
		Schist	
		Gneiss (pronounced 'nice')	
	Basalt	Eclogite	• Very dense rock from the upper mantle • Large crystals of garnet surrounded by green pyroxene
		Amphibolite	• Dark • Dense • Poorly foliated • Rich in hornblende and plagioclase feldspar
	Granite	Gneiss	• Hard • Dense • Bands (may be a centimetre or more across) of dark biotite mica and light feldspar/quartz • There may be other metamorphic minerals growing in the layers.
	Limestone	Marble	• Granular (large crystals) • Dense • Hard, although grains are soft

6.5 LEARNING CHECK

1 **Identify** the type(s) of rocks from which metamorphic rocks form.

2 **List** the conditions in which contact metamorphic rocks form.

3 **Describe** the conditions under which regional metamorphic rocks form.

4 **List** four properties of metamorphic rocks.

5 **Compare** the features of a metamorphic rock formed by heat with one formed by both heat and pressure.

Video activity
Rock cycle

Interactive resource
Label: The rock cycle

Extra science investigation
Modelling the rock cycle

rock cycle
a model used to explain how rocks are formed and how they change

GET THINKING

Look at Figure 6.6.1. What does it show? How does it relate different rock types to each other?

What is the rock cycle?

Scientists use models to understand the world. A model geologists use to explain how rocks are formed and change is called the **rock cycle**. Consider the diagram of the rock cycle shown in Figure 6.6.1.

The rock cycle shows the processes through which one rock type or material can be changed into another. It is called a cycle because it is possible to move from one substance to another and eventually return to the start.

Energy changes in the rock cycle

Energy is necessary for change to occur. Heat is needed to melt rocks. A loss of heat causes igneous rocks to crystallise. Energy is needed to produce the pressure for

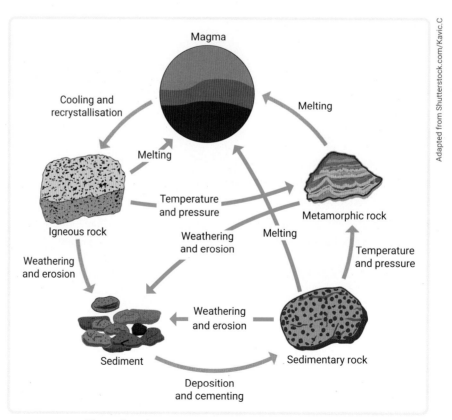

▲ **FIGURE 6.6.1** The rock cycle

metamorphism and to alter sediment as it moves down a river. The heating and cooling of rocks can also cause weathering. Look at the rock cycle (Figure 6.6.1) and see whether you can identify the source of energy needed for the changes that occur.

Even rocks that have been formed deep under the surface of Earth can eventually be exposed. The process whereby underground rocks are raised to the surface is called **uplift** (Figure 6.6.4). As rocks weather and are removed from the surface by erosion, deeper rocks become exposed.

▲ **FIGURE 6.6.2** This lava flow at Fagradalsfjall in Iceland slowed down as it went down the slope. As it slowed, it cooled and became solid rock.

▲ **FIGURE 6.6.3** Wind and waves have movement energy and they can cause weathering and erosion.

▲ **FIGURE 6.6.4** Uplift, weathering and erosion have formed the Bungle Bungle Range in Western Australia.

uplift
the process whereby rocks formed underground are raised to the surface

6.6 LEARNING CHECK

1 Draw the rock cycle. Next to each rock type, write the names of two examples.
2 **Describe** the two processes required to turn a metamorphic rock into an igneous rock.
3 Make a list of the steps needed to turn a metamorphic rock into a sedimentary rock.
4 How does an igneous rock become a metamorphic rock?
5 At Broken Hill in New South Wales, there are surface rocks that formed more than 20 kilometres below the surface.
   a **Predict** what type of rocks they are.
   b **Explain**, using the rock cycle, how those rocks have been exposed at Earth's surface.

6.7 Fossils

✓ describe the conditions under which fossils form
✓ explain how the age of fossils can be determined.

▲ FIGURE 6.7.1 An example of a fossil

GET THINKING

What do you see in the photo in Figure 6.7.1? What is it made of? Was the object once living? If so, was it an animal or a plant?

What is a fossil?

Fossils are the remains or traces of organisms that one lived on Earth, preserved in rock (see Figures 6.7.4 and 6.75 on page 168). Fossils formed from the remains of plants or animals are called **body fossils**. Many living things have **hard parts** such as teeth, shells and wood. These hard parts are more likely to be preserved as fossils than soft parts of animals or plants, because they are less likely to break down or lose their shape. However, soft-bodied animals have left impressions in ancient rocks. Insects trapped in amber (the hard resin of ancient trees) have provided important information about how insects have changed over time. Sometimes, a metamorphic rock may contain fossils if the rock is not changed very much.

▲ FIGURE 6.7.2 A dinosaur footprint is a trace fossil.

Footprints, body impressions, fossilised faeces and preserved burrows are also fossils. Although these are not formed by the actual body remains of a living thing, they are structures or impressions left by the living thing. These are called **trace fossils** (Figure 6.7.2). They tell us how animals lived and moved even when the animal's remains are not present. Carvings and other objects made by humans are called **artefacts** and they can sometimes be confused with fossils.

fossil
the remains or traces of living organisms, preserved in rock

body fossil
a fossil formed from the remains of a plant or an animal

hard parts
refers to the hard parts of an organism's body, such as shells, bones and teeth

trace fossil
a structure or an impression left by a plant or animal, which shows that life existed

artefact
an object made by a human being

〰 Going further: how big was that dinosaur?

Scientists have developed some rules to work out the size of a dinosaur from its footprint. They:

• measure the length of the footprint
• multiply the length of the footprint by 4 to estimate the height of the dinosaur's hip
• multiply the length of the footprint by 10 to estimate the head-to-tail length of the dinosaur.

You try it. Use the hand in Figure 6.7.2 to estimate the length of the footprint in centimetres, then work out the other lengths.

How fossils form

Figure 6.7.3 shows the steps involved in fossil formation.

Remains

The organism is dead and its remains settle on the surface (underwater or on land).

Rapid burial

The remains of a living thing must be buried quickly to avoid being eaten or decaying. Burial in fine sediments, such as those found on the sea floor, is one of the best environments in which to form a fossil.

A condition such as rapid burial is unlikely in some environments, which means that plants and animals living in certain areas may never have become fossils.

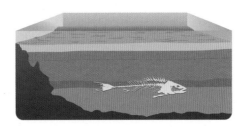

Lithification

The sediments turn into rock due to compaction. This process is called lithification.

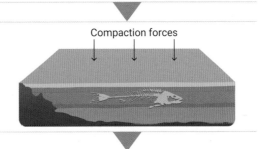

lithification
the process whereby a sediment under compaction becomes a rock

Alteration

In most fossils, the hard parts are altered. Sometimes the remains dissolve and a cavity is left in the rock. This is called a mould. The mould may fill with new minerals, which may replace shell or fill pores in wood, creating a cast of the original organism.

Sometimes heat changes the remains into carbon, which leaves a black impression of the organism.

mould
a fossil made in the shape of a plant or an animal's remains after those remains have dissolved

cast
a fossil formed when minerals fill a fossil mould

Uplift and erosion

Finally, for humans to find the fossil, it must be moved from the place where it formed to the surface of Earth.

▲ **FIGURE 6.7.3** Fossil formation

▲ FIGURE 6.7.4 A cast (left) and mould (right) of an ammonite fossil

▲ FIGURE 6.7.5 A fossil of dinosaur bones. Most of the bone has been replaced by minerals.

☆ ACTIVITY

Modelling fossils

You need

- a leaf or shell
- plasticine or playdough
- disposable cup
- plaster powder
- water
- pop stick for stirring

What to do

1 Mould the plasticine into a rectangular prism or cube shape. Press a leaf or shell into the plasticine, then carefully remove it. You have just created a mould.

2 Put some plaster powder in a disposable cup and gradually mix in small amounts of water using the pop stick until the mixture forms a thick paste. Pour the paste into your mould.

3 Leave the plaster to set.

4 Carefully remove the plaster and observe the cast you have made.

What do you think?

1 How does the cast compare with the mould?

2 Did the process cause any distortions in the cast? Do you think this might happen in nature as a fossil forms?

Determining the age of fossils and rock layers

The age of a fossil can be determined using the order of rock layers in which it is found. We know that the oldest sedimentary layers are at the bottom and the youngest are at the top of the rock.

Putting fossils into an order from youngest to oldest is called **relative dating**. Relative dating does not require us to put a date on something. It is about the order in which events occur. The use of **radioactive elements** in minerals to put an actual date on an event is called **absolute dating**. Radioactive elements are part of mineral crystals in igneous rocks. Over time, the radioactive atoms disintegrate and change into other atoms. Scientists know this and can use this to work out how old a rock is (Figure 6.7.6).

relative dating
putting fossils, rocks or events into the order in which they occurred

radioactive element
an element that can produce radioactivity

absolute dating
using scientific equipment to determine how old something is

9780170463027

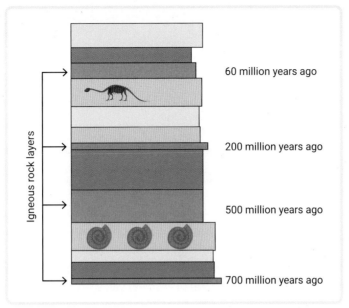

60 million years ago

200 million years ago

500 million years ago

700 million years ago

Igneous rock layers

▲ **FIGURE 6.7.6** Scientists can work out the relative age of a rock or fossil by looking at the order of rock layers. However, to find out the absolute age of a rock or fossil, scientists use the elements in minerals in igneous rocks.

Video activity
How do fossils form?

Interactive resource
Label: Fossil formation steps

6.7 LEARNING CHECK

1 What is a fossil?

2 In what types of rocks do fossils form?

3 What sorts of things can become a fossil?

4 Figure 6.7.7 shows layers of sedimentary rocks from three locations. Fossils found in the rocks are represented by different symbols.

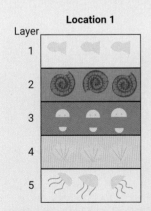
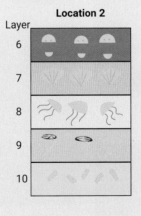
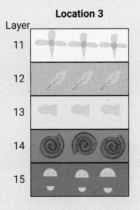

Location 1

Layer
1
2
3
4
5

Location 2

Layer
6
7
8
9
10

Location 3

Layer
11
12
13
14
15

▲ **FIGURE 6.7.7**

a Which is the oldest layer within the three locations?

b Draw the youngest fossil shown in the sections.

5 Make a flow chart showing the stages in creating a fossil found in a rock.

6.8 First Nations Australians' shaping of rocks into tools

IN THIS MODULE, YOU WILL:
- ✓ recognise that First Nations Australians have long quarried and mined rocks and minerals
- ✓ examine how First Nations Australians applied their traditional geological knowledge to select and process rocks for different purposes.

Important!

Please remember to never touch or disturb First Nations Australian artefacts. Report any finds to your local cultural heritage authority.

Obtaining rocks and minerals: quarrying and mining

First Nations Australians have a deep understanding of the natural environment and its materials. They also have great knowledge and skills to design tools that are flexible, adaptable and sometimes very specific to a certain task. These factors are the key to why their technologies have been so successful over many years.

For many tens of thousands of years, First Nations Australians have been mining and quarrying the land for a variety of rock and ochre types. In the eastern states alone, there are several hundred recorded First Nations Australians' rock and mineral extraction sites. At Wilgie Mia, in Western Australia, the Wajarri Yamatji Traditional Owners have mined ochre for an estimated 30 000 years, making it the world's oldest continuous mining operation.

▲ FIGURE 6.8.1 First Nations Australians' sandstone quarry on Mithaka Country, far western Queensland

▲ FIGURE 6.8.2 Debris from silcrete quarry and tool making, Mithaka Country

◄ FIGURE 6.8.3 (a) Evidence of reduced sandstone blank for the production of a grindstone, Mithaka Country; (b) completed grindstone, Mithaka country. Note peck marks from the shaping process.

The locations of valuable rock and mineral deposits were well known to First Nations Australians and ownership of the mine or quarry rested with the cultural group on whose land it was located. Although most stone tools could be made from readily available local materials, highly specialised tools, such as those made from Mt William greenstone, basalt and natural glass, were highly prized and important for trade between groups.

1 **Discuss** why Mt William greenstone axes have traditionally been found to be used by many different First Nations Australians groups.

☆ **ACTIVITY 1**

Manufacturing artefacts: from rocks to tools

Through observation and trial and error, First Nations Australians developed a deep understanding of the properties of different rocks and minerals. They applied this understanding to make stone tools for different purposes, including food processing, cutting, hunting, scraping and cleaning animal skins, and to make other tools. Ochre minerals were processed to produce pigments used for cosmetics, body and artefact decoration and visual representations (cultural paintings).

First Nation's Australians achieved two world firsts with stone technology. They were the first to produce ground edges on cutting tools and the first to grind seeds.

Chopping tools such as hatchets (a small axe attached to a handle) were used for a variety of reasons, including to fell trees, to remove sheets of bark for housing and canoes, to cut shields out of trees and to cut toe holds in trees for climbing. Hatchets needed to be made from strong, hard rock that did not fracture under high impact, and which could also be ground to produce a wedge with a sharp edge. The main type of rock used to make these chopping tools was igneous rock, such as fine-grained basalt.

Cutting and scraping tools – such as flake and scraper knives, spearheads and chisels – were made from rock that could be flaked to produce a sharp edge. Rock types commonly used were hard sedimentary rock such as chert and flint, and metamorphic rock such as quartzite.

▲ **FIGURE 6.8.4** A selection of First Nations Australian stone axes (origin unknown)

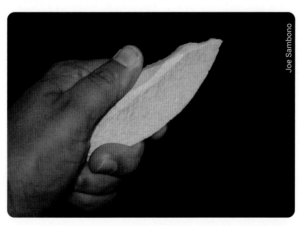

◄ **FIGURE 6.8.5** First Nations Australian stone knives (origin unknown) with spinifex resin handle

▲ **FIGURE 6.8.6** A Nardoo grindstone. The top stone is used for pounding and grinding Nardoo sporocarps (the fruit bodies of an aquatic fern), Mithaka Country.

First Nations Australians used pounding and grinding tools for processing food such as grinding seeds and nuts, or processing ochre. Usually, sedimentary rocks such as sandstone and metamorphic rocks such as quartzite and schist were used to manufacture grindstones (also called millstones by archaeologists) because the surface was more abrasive. Grindstones were often shaped to produce a large depression in the surface. Substances to be ground were placed in the depression and a round or an oval top stone (archaeologists call this a muller stone) was used to finely grind these substances.

☆ ACTIVITY 2

Testing rocks for hardness and strength

In this activity, you will test the hardness and strength (resistance to breaking or being deformed) of rock samples to determine the best type for making a hatchet.

You need

- samples of different types of rock (igneous, sedimentary and metamorphic)
- steel nail
- hammer
- hard surface such as a concrete path

> ⚠ **Warning**
>
> Wear safety glasses and protective clothing to avoid being injured by flying pieces of rock. Hold the hammer firmly when hitting the rock to avoid it slipping from your grip.

What to do

1 Examine the rock samples provided and predict which would make the best hatchet head.

2 Construct a results table to record the type of rock, effect of scratching with a nail and effect of being hit with a hammer.

3 Using firm pressure, draw the sharp end of a nail over the surface of the rock sample. Record the results.

4 Repeat step 3 for all samples.

5 Place a rock sample on a hard surface (such as a concrete path) and firmly strike it with a hammer. If there is no effect, repeat until an effect is observed, to a maximum of four blows. Record the results.

What do you think?

1 **Analyse** the results and discuss which sample of rock would make the best hatchet head.

2 **Compare** this to your prediction. Explain why they match or don't match.

3 Was this a fair test? Explain your answer.

6.9 Dealing with acid mine water

BY THE END OF THIS MODULE, YOU WILL BE ABLE TO:

✓ explain why mines need to be carefully rehabilitated to prevent damage to creeks and waterways.

Acid mine drainage

Figure 6.9.1 shows a river in Spain called the Rio Tinto. There is little life in the river and its colour comes from minerals dissolved in the acidic river water. How did the river come to be like this, and how can we prevent similar fates for our waterways?

Copper is an example of a valuable resource that we extract from the earth. The most commonly mined copper minerals are chalcopyrite ($CuFeS_2$) and chalcocite (Cu_2S). These are called sulfides because the metals are combined with the element sulfur. The Rio Tinto River drains an area where a huge amount of sulfide ore was located.

▲ **FIGURE 6.9.1** The Rio Tinto River in Spain

Acid mine drainage is an environmental problem. It is caused when sulfide minerals contact oxygen in the air. The minerals weather, producing compounds of sulfur and oxygen called sulfur oxides. When these oxides dissolve, they react with water and form acids. The acidic water will damage life in rivers and creeks if it escapes out of the mine. The acid mine water is also capable of dissolving elements such as copper, zinc, arsenic, cadmium and lead. These metals can end up in plants and animals and, over time, may cause illness and death.

Mine rehabilitation is the process undertaken by scientists and engineers to clean up a mining area. Mining can cause damage to the environment, so it is important to clean up or rehabilitate mining sites after they are closed. Mine clean-up is incredibly important to reduce the impact that mining has on the environment. For example, it involves treating any acidic water the mine has produced so that rivers are not polluted. The cost of treating acid mine drainage is high but the cost of damaging the environment is even higher.

Video activity
Cultural reconnection:
Mine rehabilitation

6.9 LEARNING CHECK

1 **Describe** the steps in which acidic, metal-rich mine water is produced/formed.
2 **Explain** why drainage from acid mines should not be allowed to enter rivers or streams.
3 The Rio Tinto River receives water from areas that have been mined for 5000 years. In previous times, why did the miners not take steps to prevent the discharge of acid mine waters?

SCIENCE INVESTIGATIONS

6.10 Fieldwork

SCIENCE SKILLS IN FOCUS

IN THIS MODULE, YOU WILL FOCUS ON
LEARNING AND IMPROVING THESE SKILLS:

▶ form predictions based on observations
and the use of scientific models

▶ plan and conduct procedures to collect
data

▶ process, model and analyse data

▶ consult with Aboriginal and Torres
Strait Islander Peoples to conduct
investigations on Country/Place

▶ describe a range of minerals using
their physical properties.

Fieldwork allows you to learn more about the
environment in which you live. It is important
to make minimal changes to the environment
while you are doing
fieldwork, and to care for
the land. One important
aspect of caring for
the environment is
recognising First Nations
Australians' artefacts
and heritage sites of
significance, and ensuring
you cause no harm to
those sites, while showing
respect for Country/Place.

Video
Science skills
in a minute:
Representing
data

**Science skills
resource**
Science skills
in practice:
Representations
of data

Here are some things to consider in planning your
fieldwork.

▶ **Preparation prior to entering the field**

• Whose Country/Place will you be visiting? Can
you consult with First Nations Australian Elders
about your visit? How will you get in touch with
First Nations Australian Elders in your area?
Have you asked a First Nations Australian Elder
if they would spend some time showing you the
area?

• Questions to be asked of First Nations
Australians custodians: Are there areas you
need to avoid or be careful of when working
on Country/Place in a particular way? Are
there any particular artefacts you might find on
the field trip? What should you do if you find
a First Nations Australians' artefact on your
trip? Is there advice on any hazards you might
encounter?

▶ **Researching artefacts**

• What do First Nations Australians' stone
artefacts look like? Sometimes, artefacts are
made of rocks not found in your local area.

• Your state/territory museum is likely to have
online images of First Nations Australians'
artefacts. Make yourself familiar with the
appearance of stone tools. You could also refer
back to Module 6.8.

▶ **Dealing with outcrops, artefacts and
heritage sites encountered on the field trip**

• Do not touch or disturb artefacts. How might you
record what you find? Who will you tell about
your find?

• Do not break rocks from outcrops. If samples
are to be collected, only take small amounts and
select loose samples on the ground. It is best
to photograph, draw and describe rocks and
outcrops while you are in the field.

• If you find a heritage site, leave it untouched and
move away to another area in the Country/Place
you are on. Let others know, so they can avoid the
site too.

• If you encounter First Nations Australians'
artwork, do not touch it. This damages the
artwork.

9780170463027

INVESTIGATION 1: PLANNING TO CONDUCT FIELDWORK ON COUNTRY/PLACE

AIM

To prepare a guide for procedures to be followed when encountering First Nations Australians' artefacts and heritage sites on field trips

YOU NEED

☑ two or three other students with whom to discuss ideas and information

WHAT TO DO

1 Use the information in Science skills in focus to make a list of points to include in the guide for your field trip. Start by identifying where you can find help from First Nations Australians and the questions you wish to ask them.

2 With the assistance of your teacher, see whether a First Nations Australian Elder or their delegate can answer your questions.

3 Research First Nations Australians' stone tools and other artefacts you might encounter. Make drawings or descriptions to help you identify these products if you encounter them on your field trip.

4 Research commonwealth, state and territory cultural heritage laws that must be observed when working on Country/Place.

5 Use the information about laws, dealing with outcrops and artefacts to make a set of rules students should follow on the field trip. Write a reason for each rule.

6 Assemble your information into a one- or two-page handout as a reference for the field trip.

WHAT DO YOU THINK?

1 Why should we develop procedures to follow when we encounter First Nations Australians' artefacts or heritage sites?

2 What did you learn during this investigation about the history of First Nations Australians in your area?

3 How confident are you that you can identify artefacts, and that you will know what to do if you find them?

CONCLUSION

Have you made a useful guide to encountering artefacts and heritage sites on your fieldtrip?

INVESTIGATION 2: MINERAL RESEARCH

AIM

To describe a range of minerals using their physical properties

YOU NEED

- ☑ five different mineral specimens
- ☑ hand lens or magnifying glass
- ☑ Mohs scale of hardness test kit
- ☑ unglazed ceramic plate

WHAT TO DO

1 Draw up a table to record your results. It should include six rows (one for the column headings and one for each mineral specimen) and five columns with the headings: 'Specimen', 'Colour', 'Hardness', 'Cleavage and fracture' and 'Special features'.

2 Carefully record the colour of each mineral. Note any variations in colour you see.

3 Use the Mohs hardness test kit to work out the hardness of each mineral. Do this by trying first to scratch the fluorite test mineral with your mineral specimen. If it does not scratch the fluorite, try the calcite (hardness 3). If it does scratch the fluorite (hardness 4), try the apatite (5) and then orthoclase (6) until you find a mineral your sample cannot scratch. Test each sample until you work out the hardness range (harder than ..., but softer than ...) Record the hardness as a single number or a range in the results table.

4 Cleavage or parting surfaces are flat or appear as parallel lines on the surface of the mineral. Try to identify the cleavage planes and the angles between them. If a mineral breaks unevenly and not along a flat surface, it is called a fracture. Record what you find in the table.

Warning

Beware of small pieces of minerals entering your eyes. Beware of sharp edges on rock specimens. There may be toxic chemicals in the specimens. Wear gloves and safety glasses.

5 Look carefully at the specimen. Is there anything that stands out? Does the mineral have a characteristic crystal shape? Does it look like a metal? Does it feel heavy? Record what you see and feel.

RESULTS

Make sure your table is complete and has a title.

WHAT DO YOU THINK?

1 Identify the property that was the easiest to determine.

2 Did any minerals have the same properties?

3 Which mineral was the hardest? Which was the softest?

4 What was the most difficult property to work out? Why do you think this was so?

CONCLUSION

Did the properties allow the different minerals to be described? Write a conclusion that summarises your findings. Make sure it relates to your aim.

9780170463027

1 **Identify** the features/properties of a mineral.

2 What term is used for the following mineral properties?
   a How easy it scratches
   b Splitting along flat surfaces
   c The way a surface reflects light
   d The colour of powdered mineral

3 State an example of each of the following rock types.
   a An igneous extrusive rock
   b A volcanic glass
   c A sedimentary rock made up of clay particles
   d A contact metamorphic rock
   e A regional metamorphic rock

5 **Identify** whether the following sentences are true or false. If the statement is false, rewrite the sentence to make it true.
   a Minerals are composed of two or more chemical elements.
   b Granite is an example of an intrusive rock.
   c Weathering is the chemical breakdown of minerals.
   d Heat and pressure are the causes of metamorphic change.
   e Absolute dating provides an age for something in numbers of years.
   f Trace fossil is the name given to the preserved soft parts of an animal or plant.

4 Copy and complete the table by placing a tick for each rock type with the property listed.

Property	Rock type		
	Igneous	Sedimentary	Metamorphic
Composed of interlocking mineral crystals			
May contain large crystals			
Has a texture called foliation			
May contain gas bubbles			
Contains fragments and particles cemented together			
May contain fossils			

6 **Explain** why sediments at the start of a river are different from those found where the river enters the ocean.

7 **List** the properties you would look for to distinguish metamorphic rocks from igneous rocks.

8 Why is a shellfish more likely to become a fossil than a jellyfish?

9 What process will turn a metamorphic rock into magma?

10 **Explain** the difference between ore and a rock.

11 **Explain** why sand blown from a beach into sand dunes has a finer grain size than the sand on the beach.

12 Volcanic ash is composed of shiny, sharp fragments of volcanic glass. What does this tell you about the cooling of the magma that formed the ash?

13 Use your understanding of the rock cycle to **create** a flow chart showing how a sediment can become a metamorphic rock such as slate.

14 **Justify** why an animal sinking into a swamp is more likely to be fossilised than the same animal sinking into a river.

15 **Describe** the properties of a sediment that would help you identify whether it had been transported a long way from where it was eroded.

16 Why can a metamorphic rock have bedding and foliation but a sedimentary rock can only show bedding?

17 a **Identify** one type of stone tool constructed and used by First Nations Australians.

 b **Describe** the properties of rock required for this tool and **explain** how these related to the use of the tool.

 c **Name** one type of rock used in constructing this tool.

18 What features would help you distinguish between a cast fossil and a mould fossil?

19 Light-coloured minerals such as quartz and feldspar are more common in beach sand than dark-coloured minerals such as pyroxene and olivine.

 a What does this suggest about the differences in properties of the light- and dark-coloured minerals?

 b **Explain** why beaches on the volcanic island of Hawai'i contain lots of dark minerals such as olivine.

20 If you dig a deep hole in soil, the soil becomes harder and harder to dig out of the hole. Why does the soil become harder to dig and what would have to happen to turn the soil into a rock?

21 **Explain** how pressure and temperature lead to the characteristics, features or properties of regional metamorphic rocks.

22 **Explain** why diamond and graphite, both composed of carbon, are classified as different minerals.

23 The table contains a list of properties for five different minerals.

Mineral	Colour	Hardness	Cleavage	Lustre	Density
Olivine	Green-brown	6.5	None	Glassy	3.2–3.4
Orthoclase	White, pink, light green, brown	6	2 at 90°	Glassy	2.6
Quartz	Colourless or a wide range of colours	7	None	Glassy	2.6–2.7
Biotite	Black to dark brown	2.5–3	1	Pearly	2.8–3.4
Calcite	Colourless or a wide range of colours	3	3	Glassy to pearly	2.7

 a Which mineral will be broken down most quickly in a swiftly flowing river?

 b Which properties would be the best ones for distinguishing olivine and biotite from the other minerals?

24 Basalt weathers to form clay-rich and fertile soils. Granite weathers to form sandy, low-fertility soils. **Explain** why these igneous rocks give rise to such different soils.

25 **Create** a concept map of the key terms in this chapter. Make links between the terms and label the links to show what the relationships are.

BIG SCIENCE CHALLENGE PROJECT #6

1 Connect what you have learned

In this chapter you have learned about the characteristics/ properties/features of different types of rocks and the processes that change them. Examine the image on this page carefully and make a list of hypotheses about the possible rock types and processes operating on Mars.

2 Check your thinking

The rock at the front of the picture seems to contain holes. It might be a volcanic rock with gas bubbles. It might be a sedimentary rock with parts that have weathered and eroded away. What features of the rock would you look for to decide what type of rock it is?

Alamy Stock Photo/NASA Image Collection

3 Make an action plan

Make a plan to research the types of rocks found on Mars. The largest volcano in our solar system is on Mars. Find out what types of rocks are produced by volcanoes there.

Rocks in the image appear to be sedimentary. Are sedimentary rocks common on Mars? If so, what transported the sediment?

4 Communicate

Create a modified diagram of the rock cycle for Mars. Include the rock types and processes you have inferred or researched.

7

Plate tectonics

7.1 **Structure of Earth** (p. 182)

Earth is composed of layers with different compositions and properties.

7.2 **Plate tectonics** (p. 184)

The plate tectonics model explains how plates create the surface features of Earth.

7.3 **Plate boundaries** (p. 188)

Plates interact with each other in different ways.

7.4 **Continental drift** (p. 192)

Continents move over the surface of Earth.

7.5 **Seafloor spreading** (p. 194)

Oceans are formed by a process called seafloor spreading.

7.6 **Earthquakes and volcanoes** (p. 198)

Earthquakes, volcanoes and tsunamis are generated by tectonic processes.

7.7 FIRST NATIONS SCIENCE CONTEXTS: **First Nations Australians' records of geological events in Australia** (p. 202)

Knowledge of geological events has been passed from generation to generation of First Nations Australians through their oral traditions.

7.8 SCIENCE AS A HUMAN ENDEAVOUR: **Discovering the ocean floor** (p. 204)

Our understanding of plate tectonics and how the seafloor forms and changes has evolved over a very long time.

7.9 SCIENCE INVESTIGATIONS: **Researching and presenting scientific data** (p. 206)

1 Using maps to summarise information and show patterns
2 Volcanic eruptions
3 Tsunamis and earthquakes
4 Cheesy plate tectonics

SOUTHERN Biological

BIG SCIENCE CHALLENGE

▲ **FIGURE 7.0.1** A volcanic landscape in Iceland

In some parts of the world, such as Iceland, it is easy to see Earth's surface being shaped by enormous forces.

What do you see in Figure 7.0.1? Why do the hills in the distance have a cone shape? Why is the valley so broad and flat? Why is the ground composed of dark rocks?

▶ **Why is the Icelandic landscape so different from that of Australia?**

#7 SCIENCE CHALLENGE ACCEPTED!

At the end of this chapter, you can complete Big Science Challenge Project #7. You can use the information you learn in this chapter to complete the project.

Assessments
- Prior knowledge quiz
- Chapter review questions
- End-of-chapter test
- Portfolio assessment task: Data analysis

Videos
- Science skills in a minute: Presenting data in different forms **(7.9)**
- Video activities: Structure of Earth **(7.1)**; Plate tectonics **(7.2)**; The scientist behind the plate tectonics theory **(7.4)**; What is a volcano? **(7.6)**; What is an earthquake? **(7.6)**; Mapping the secrets of the ocean floor **(7.8)**

Science skills resources
- Science skills in practice: Data in different forms **(7.9)**
- Extra science investigations: Modelling seafloor spreading **(7.5)**; Modelling P and S waves **(7.6)**

Interactive resources
- Label: Earth's structure **(7.1)**; Features of plate boundaries **(7.3)**; Features of the seafloor **(7.5)**
- Drag and drop: Convection currents and slab pull **(7.2)**
- Match: Types of boundaries **(7.3)**

❖ Nelson MindTap

To access these resources and many more, visit:
cengage.com.au/nelsonmindtap

BY THE END OF THIS MODULE, YOU WILL BE ABLE TO:

✓ describe the layered structure of Earth.

Video activity
Structure of Earth

Interactive resource
Label: Earth's structure

geosphere
the solid part of Earth

atmosphere
the gaseous layer surrounding Earth

GET THINKING

How many layers make up Earth? Which is the largest? Survey the module and see whether you can count how many layers there are.

Earth's structure

Earth is composed of four spheres that interact with each other to form the 'Earth system'. The solid Earth is referred to as the **geosphere**. The gaseous envelope surrounding the Earth is the **atmosphere**, and it is made up of many different layers (Figure 7.1.1). The atmosphere contains the air in the Earth system and is held by Earth's gravity. The atmosphere is mainly composed of nitrogen gas and oxygen gas. The atmosphere extends thousands of kilometres above the surface; however, three-quarters of all the gas in the atmosphere is in the lowest 11 kilometres.

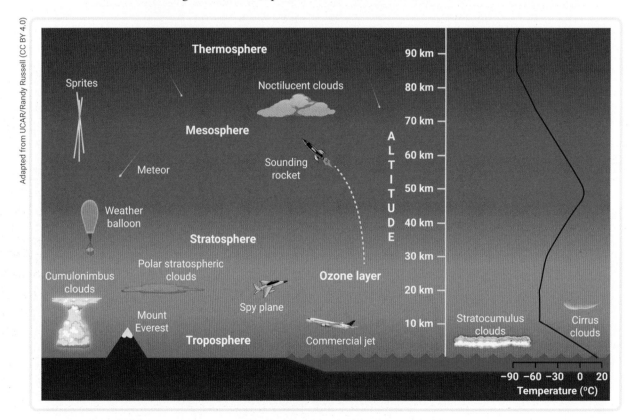

Adapted from UCAR/Randy Russell (CC BY 4.0)

▲ **FIGURE 7.1.1** Earth's atmosphere is made up of many layers.

hydrosphere
all the water on Earth

All of the solid, liquid and gaseous water on Earth forms the **hydrosphere**. Most of the hydrosphere is made up of rivers, lakes and oceans, but ice sheets, water vapour in the atmosphere and water in the rocks of Earth's crust are also considered to be part of the hydrosphere. We consider our oceans to be vast and very deep, but the deepest part of any of Earth's oceans – the Mariana Trench in the Pacific Ocean – is only 11 kilometres

9780170463027

deep (Figure 7.1.2). Earth's radius is 6371 kilometres, more than 500 times that depth.

Life occurs in all the outer layers of Earth, including deep in the crust. All the places where life is found comprise the **biosphere**.

The geosphere

The geosphere is divided into four layers: the **crust**, **mantle**, **outer core** and **inner core**. The layers differ in composition and whether they are solid or liquid (Figure 7.1.3). With increasing depth, the temperature and pressure in Earth's layers increases. The mantle's temperature ranges from 200°C to 4000°C. The surface temperature of the inner core is similar to the surface temperature of the Sun – about 5400°C.

The crust is the outermost and thinnest layer of Earth. The crust can be divided into the oceanic and continental crust. Oceanic crust forms under oceans and is less than 10 kilometres thick. It is dense and is generally composed of the igneous rocks basalt and gabbro. Continental crust is thicker than oceanic crust and forms the continents and nearby shallow oceans. Continental crust is composed of a variety of rocks and is much older than oceanic crust.

The mantle lies between the crust and the core. It makes up two-thirds of the solid Earth's mass and 84 per cent of the volume of Earth. The mantle is made of iron-rich rock and is solid, but some sections of it are considered to have plasticity, meaning that the mantle moves very slowly over time. Over millions of years, mantle flows carry heat between the core and the crust.

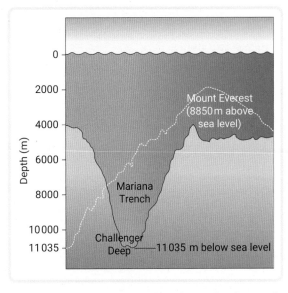

▲ **FIGURE 7.1.2** The Mariana Trench is 11 km deep and Mount Everest is almost 9 km above sea level; these are small distances compared to the radius of Earth.

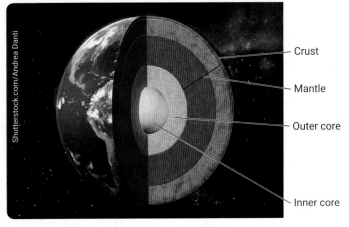

▲ **FIGURE 7.1.3** The layered structure of Earth

The core of Earth is composed of iron and nickel metal. It is divided into the outer and inner core. The outer core is liquid and the inner core is solid. This difference in state is caused by enormous pressure deep in Earth that changes the melting point of the metals found in the inner core. The pressure and density are so high that it causes the inner core to be solid. Earth's magnetic field is created by movement of liquid materials in the outer core.

biosphere
the parts of Earth where life is found

crust
the outermost and thinnest layer of Earth

mantle
the rock layer between the crust and the core of Earth

outer core
the liquid part of Earth's metallic core

inner core
the innermost solid, metal part of Earth

7.1 LEARNING CHECK

1 **List** the layers of Earth that are liquid, gaseous or solid.
2 **Identify** the thickest layer of the solid Earth.
3 Make a table of the layers of the geosphere including the layer name, what it is made of and its thickness. You may need to do some extra research.
4 **Contrast** ocean and continental crust, identifying as many differences as possible.
5 Draw a labelled diagram of the structure of Earth.

BY THE END OF THIS MODULE, YOU WILL BE ABLE TO:
✓ state the theory of plate tectonics
✓ describe the processes that cause Earth's plates to move.

GET THINKING

Look closely at the illustrations in this module. What do you think they are trying to show?

lithosphere
Earth's crust and the rock-like upper part of the mantle

asthenosphere
the solid part of the mantle below the lithosphere that can flow

The theory of plate tectonics

Two important components of the geosphere are the **lithosphere** and **asthenosphere** (Figure 7.2.1). The lithosphere is made up of Earth's crust and the brittle, rock-like part of the mantle underneath it. Forces can cause the lithosphere to change shape or fracture. The lithosphere is thickest under continents and thinnest under oceans.

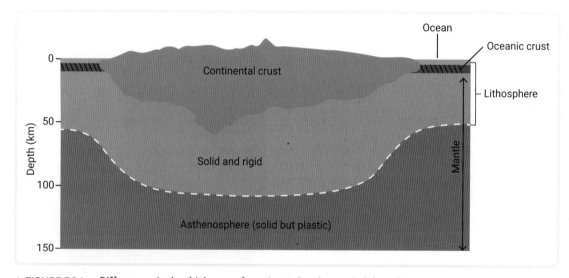

▲ **FIGURE 7.2.1**   Differences in the thickness of continental and oceanic lithosphere

Video activity
Plate tectonics

Interactive resource
Drag and drop:
Convection currents
and slab pull

The plate tectonics theory states that Earth's lithosphere comprises pieces called plates that move relative to each other. The movement of the plates causes the major topographic features of Earth: mountains (on land and under the ocean), ocean basins and deep-sea trenches. The word 'tectonic' comes from the Greek word *tecton*, meaning 'builder'.

Beneath the lithosphere is the asthenosphere. Although still part of the solid mantle, the asthenosphere is much hotter and more fluid than the lithosphere. On average, the asthenosphere is 180 kilometres thick. Heat from deep within Earth keeps the asthenosphere plastic so it can flow (like toothpaste) and allows the lithosphere to move. As forces act on a plate, the asthenosphere allows the plate to move horizontally and to rise or sink. When erosion removes rock from a mountain, the mountain slowly rises as the weight is removed. It is like a canoe rising in the water when someone gets out.

The plates of Earth

Figure 7.2.2 shows the major and minor plates of Earth. The arrows show the direction in which each of the plates are moving. Notice how some plates are moving away from each other and others are moving towards each other. The plates move at different speeds.

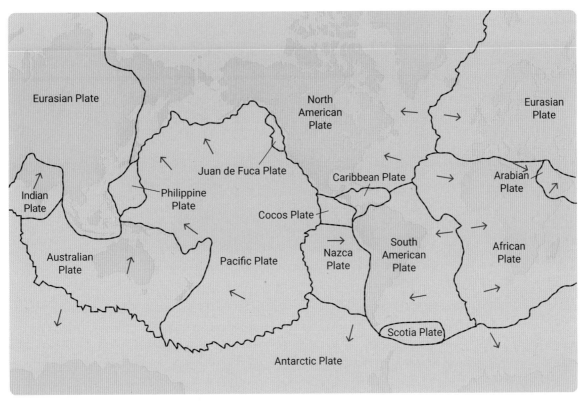

▲ **FIGURE 7.2.2**   The plates of Earth. The arrows indicate the direction the plates are moving.

There are two types of plate. An **oceanic plate** is a lithospheric plate that contains mainly oceanic crust. The Pacific and Nazca Plates are composed only of oceanic crust. A **continental plate** is a lithospheric plate containing a lot of continental crust. The Australian Plate, the African Plate and the Eurasian Plate are all known as continental plates, although they also contain some oceanic crust. Plates move at speeds ranging from less than 2 to about 20 metres per hundred years. The average rate is about 10 centimetres per year, with oceanic plates moving faster than continental plates.

Continental and oceanic crusts have different structures. Table 7.2.1 summarises the differences between continental and oceanic crust.

oceanic plate
a lithospheric tectonic plate that contains mainly oceanic crust

continental plate
a lithospheric tectonic plate containing a lot of continental crust

▼ **TABLE 7.2.1**   The features of oceanic and continental plates

Feature	Oceanic plate	Continental plate
Average thickness (km)	7	35
Average density (g/cm³)	3.0	2.7
Composition	Thin sediment layer over rocks with a composition like basalt	Wide range of rock types; composition similar to granite
Structure	Layered	Sedimentary layers and folded and forced rocks

Why plates move

ridge push
the process that moves crust away from a mid-ocean ridge

slab pull
the force driving tectonic plate motion caused by a sinking plate

convection
the process that transfers heat in a fluid

There are two main causes of plate motion: **ridge push** and **slab pull**. **Convection**, caused by temperature differences, provides the energy for plate motion. At the surface, gravity causes hot new crust to pull cold ocean lithosphere down into the mantle (Figure 7.2.3).

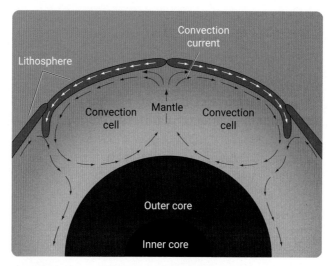

▲ **FIGURE 7.2.3** Convection and plate motion causes are related.

mid-ocean ridge
a broad, high, underwater mountain range in the ocean at a divergent plate boundary

Ridge push explains why plates move away from a **mid-ocean ridge**. Because the centre of the ridge is several kilometres above the deeper ocean floor, the lithosphere slowly slides down due to gravity. As it slides, the plate near the ridge pushes other parts of the plate ahead of it (Figure 7.2.4).

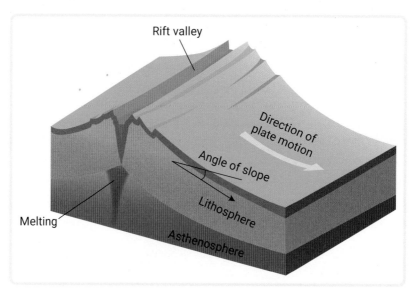

▶ **FIGURE 7.2.4**
Ridge push: lithosphere slides down the slope away from the boundary

Slab pull is caused by a sinking, or subducting, plate. It is the major cause of plate motion. When the ocean lithosphere begins to sink, it pulls the rest of the plate behind it. The plate it sinks beneath pushes up against the subducting plate and prevents the subducting plate from accelerating (Figure 7.2.5). You will learn more about how plates interact, as well as more detail about some of these terms, in the following modules.

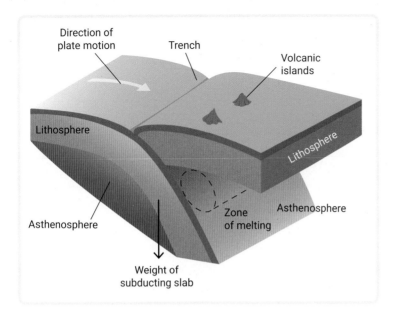

Direction of plate motion

Trench

Volcanic islands

Lithosphere

Lithosphere

Lithosphere

Asthenosphere

Asthenosphere

Zone of melting

Weight of subducting slab

◀ FIGURE 7.2.5
Slab pull: the weight of the subducting slab pulls the plate towards the trenches

Convection from a cold surface

☆ ACTIVITY

You need
- water
- 250 mL beaker
- ice cube
- food colouring

What to do

1 Add 200 mL of water to the 250 mL beaker. Wait until the water is still.

2 Carefully float the ice cube in the centre of the beaker of water.

3 Add three drops of food colouring to the water at the beaker's edge. Record what the dyed water does.

4 Carefully add three drops of food colouring to the top of the ice cube. Observe and record what happens when the food colouring enters the water.

What do you think?

1 What happens to the cooling water below the ice cube?

2 How does the movement of the food colouring support the idea of a descending body of cold fluid?

3 Why is this an example of convection?

7.2 LEARNING CHECK

1 What is a lithospheric plate?

2 How is a continental plate different from an oceanic plate?

3 **State** the theory of plate tectonics.

4 What does the theory of plate tectonics explain?

5 **Explain** ridge push.

6 **Explain** slab pull.

7 **Explain** how both ridge push and slab pull contribute to convection in the mantle.

8 Search the Internet for a copy of a blank tectonic plates map and label the plates.

Quiz
Evidence for plate tectonics

Interactive resources
Label: Features of plate boundaries
Match: Types of boundaries

convergent boundary
the border at which crust is destroyed as one plate moves beneath another, or where two continental plates collide

divergent boundary
the border at which new crust is formed as tectonic plates pull away from each other

subduction
the process in which one oceanic plate sinks beneath another plate at a convergent boundary

transform boundary
the border between two tectonic plates that are sliding past each other

---

GET THINKING

If plates move in different directions, what happens at their edges? What happens when they run into each other? Are gaps created when they move apart?

How plates interact

There are three types of plate boundaries (Figure 7.3.1). They are:

- **convergent boundaries** – crust is destroyed as one plate is pushed underneath another as they approach each other (a process called **subduction**), or two continental plates collide
- **divergent boundaries** – new crust is generated as the plates move away from each other
- **transform boundaries** – the plates slide horizontally past each other.

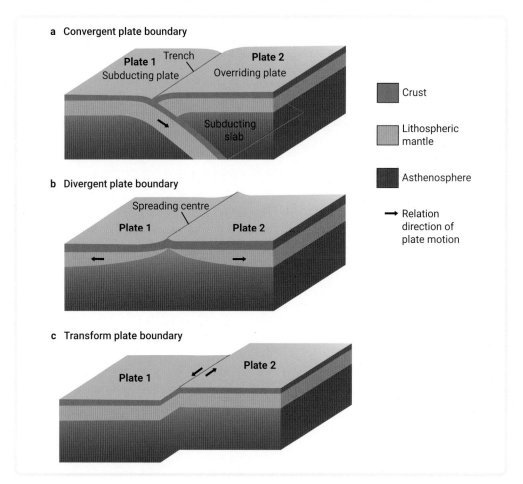

▲ **FIGURE 7.3.1** Three types of plate boundaries: **(a)** convergent, **(b)** divergent and **(c)** transform. Notice that a convergent boundary demonstrates slab pull (Figure 7.2.5, Module 7.2) and a divergent boundary demonstrates ridge push (Figure 7.2.4, Module 7.2).

As the plates move, they create distinctive structures that help to identify the processes that are happening at the plate boundary (Table 7.3.1).

▼ TABLE 7.3.1 Plate boundaries and their characteristics

Boundary type	Types of plates involved	Structures created	Volcanic activity	Earthquake activity
Divergent	Oceanic–oceanic	• Mid-ocean ridges and central rift valleys • **Faults**	Basaltic lavas from vents in the seafloor	Shallow (0– 70 km)
	Continental–continental	• Rift valleys • Faults	Volcanoes form along edges of rift	Shallow (0–70 km)
Convergent	Oceanic–oceanic	• Trench • Chain of volcanic islands • Faults	Volcanoes that erupt in an explosive way, generating large amounts of ash	Shallow to deep (0–700 km)
	Continental–oceanic	• Trench • Mountain belt with volcanoes • Faults		Shallow to deep (0–700 km)
	Continental–continental	• Fold mountains • Faults	None	Shallow to intermediate (0–300 km)
Transform	Continental–continental or Oceanic–oceanic	• Faults	None	Shallow (0–70 km)

fault
a fracture on Earth's surface where rocks have moved due to tension or compression forces

Some of the structures described in Table 7.3.1 are shown in Figure 7.3.2.

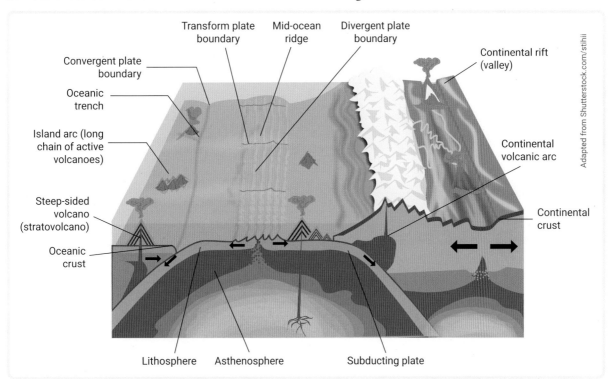

Adapted from Shutterstock.com/stihii

▲ FIGURE 7.3.2   Features of different plate boundaries

Evidence for plate tectonics

Scientists found evidence for the theory of plate tectonics during the first part of the 1900s. It built on evidence for the continental drift hypothesis (see Module 7.4).

The shape and age of the seafloor

The ocean floor was once thought to be flat and unchanging. In the 1950s, scientists found a broad, high mountain range that extended 65000 kilometres around the world in the deep ocean – a mid-ocean ridge (Figure 7.3.2). Fossils and magnetic information revealed that the seafloor was youngest at the mid-ocean ridge and got older as one moved further away from the ridge. The oldest rocks in the ocean are less than 180 million years old. This was evidence that the seafloor was indeed changeable (see also Module 7.5).

Magnetic striping of seafloor basalts

In the 1950s and 1960s, marine scientists created maps of the seafloor magnetism. The maps showed a distinctive striped pattern parallel to the mid-ocean ridge. You will learn more about this in Module 7.5. The scientists also detected transform faults – faults along a transform boundary.

Earthquake distribution

earthquake
a violent shaking of the ground caused by energy-carrying waves

Earthquakes occur where enormous forces fracture and move rocks. Earthquake activity on Earth tends to follow the edges of the tectonic plates, where the plates interact with each other (Figure 7.3.3).

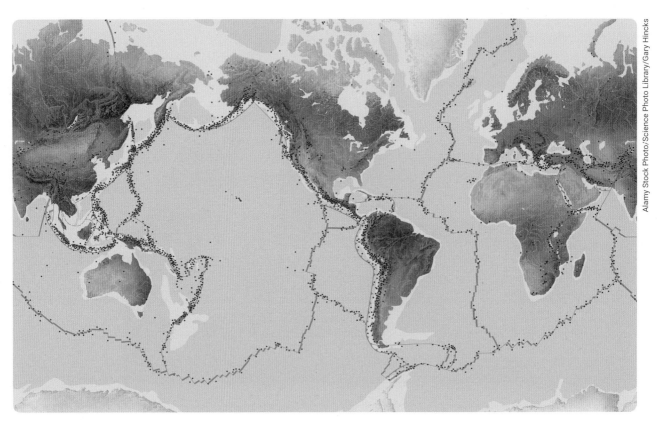

Alamy Stock Photo/Science Photo Library/Gary Hincks

▲ **FIGURE 7.3.3**　Distribution of earthquakes and volcanoes on Earth

9780170463027

Ages of hotspot volcanoes

Within the ocean there are chains, or arcs, of volcanic islands ordered in age, and created by a moving plate. As the plate carries a volcano over a **hotspot** in the mantle, the volcano stops being active and is replaced by a new one (Figure 7.3.4).

hotspot
an unusually hot area in Earth's upper mantle where the mantle melts

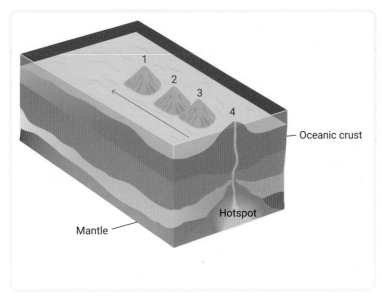

▲ **FIGURE 7.3.4** How hotspot volcanoes form. The moving plate carries volcanoes away from where they first form. Volcano 1 is the oldest. Volcano 4 is the newest.

7.3 LEARNING CHECK

1 **Describe** what occurs at a:
   a convergent boundary.
   b divergent boundary.
   c transform boundary.
2 What geographical features occur at divergent and convergent boundaries? **Explain** the processes involved in formation of these features.
3 **List** four types of evidence supporting the plate tectonics theory.
4 **Describe** two types of evidence found in the ocean for plate tectonics.

BY THE END OF THIS MODULE, YOU WILL BE ABLE TO:

✓ outline the theory of continental drift and the evidence supporting it.

supercontinent
a continent composed of all or most of Earth's continents

Pangaea
a supercontinent that once existed on Earth; the name means 'all lands'

continental drift
the movement of the continents across the surface of Earth over geological time

GET THINKING

How could all of the continents that exist today have originally been one giant continent? What evidence supports this idea?

Alfred Wegener and Pangaea

In 1912, a German scientist called Alfred Wegener proposed that in the past all the continents on Earth had been joined together as part of a **supercontinent**. Wegener named the supercontinent **Pangaea** (Figure 7.4.1).

Wegener was a meteorologist and developed his theory by studying past climates, through evidence preserved in rocks. He wondered why, 200 million years ago, there were tropical forests in the northern hemisphere, with ice sheets in the southern hemisphere at a similar distance from the equator (Figure 7.4.2).

Wegener proposed that continents could move. He found that the existence of a supercontinent could account for cold areas near the South Pole and the tropical forests near the equator. This is what we see today. Wegener's theory is known as **continental drift**.

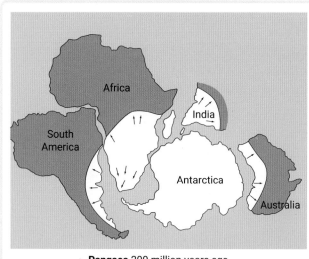

▲ **FIGURE 7.4.1** The supercontinent Pangaea

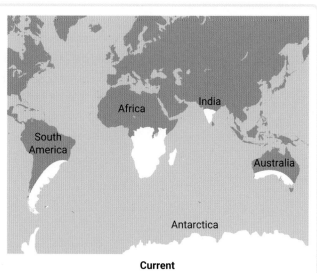

Pangaea 200 million years ago

Current

▲ **FIGURE 7.4.2** Glacial deposits in Pangaea, and where evidence of past glacial deposits is found now. Arrows show direction of ice movement.

Evidence for continental drift

Wegener assembled a lot of evidence to support his theory, including:

- the shape of continents – the coastlines of adjacent continents seem to fit together. This is best seen with the coasts of Africa and South America (Figure 7.4.2)
- rocks reflecting different climates – their distribution makes sense if they are part of a supercontinent that extended from the South Pole past the equator (Figure 7.4.3)
- fossils – animal and plant fossils would form continuous distributions across a supercontinent. It would have been difficult for land animals to travel across oceans if the continents had been separated (Figure 7.4.4)
- the continuation of geology across matching coastlines – mountain belts and old, stable parts of the continent appear to continue across coast lines at the points where continents would previously have been joined.

▲ FIGURE 7.4.3   Rock that has been polished and scratched by a glacier at Inman Valley in South Australia. This is evidence of a past glacial deposit.

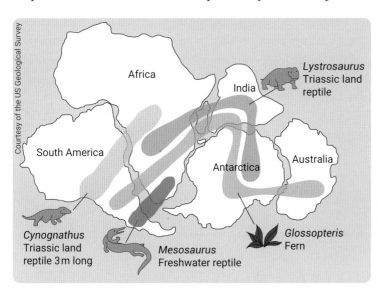

Cynognathus
Triassic land reptile 3 m long

Mesosaurus
Freshwater reptile

Glossopteris
Fern

Lystrosaurus
Triassic land reptile

Africa

India

South America

Antarctica

Australia

◀ FIGURE 7.4.4   Fossil evidence explained by continental drift

▲ FIGURE 7.4.5   *Cynognathus* was a mammal-like reptile that lived on Pangaea 251–246 million years ago.

Continental drift and convection

In Wegener's time, many scientists believed that the oceans and continents were fixed rigidly in place, and they disagreed with Wegener's theory. The biggest problem facing the idea of continental drift was that no one knew what could move a continent. A British scientist, Arthur Holmes, proposed that convection could move continents. Finally, in the 1960s, scientists discovered how oceans grow and are destroyed. (We will see how in Module 7.5.)

Video activity
The scientist behind the plate tectonics theory

7.4 LEARNING CHECK

1 **State** the theory of continental drift.
2 What was the name of the supercontinent proposed by Alfred Wegener?
3 **List** four types of evidence used to support the theory of continental drift.
4 What could Wegener's theory not explain about continental drift?
5 **Explain** how the fossil distribution of Mesosaurus (refer to Figure 7.4.4) supports the theory of continental drift.

BY THE END OF THIS MODULE, YOU WILL BE ABLE TO:

✓ describe the process of, and evidence for, seafloor spreading

✓ explain changes in the age of the ocean floor using seafloor spreading.

Quiz
Seafloor magnetic patterns

Interactive resource
Label: Features of the seafloor

Extra science investigation
Modelling seafloor spreading

GET THINKING

What is seafloor spreading? Why was its discovery an important contribution to the theory of plate tectonics?

Seafloor spreading

Until the 1950s, very little was known about the ocean floor. When US geologists Marie Tharp and Bruce Heezen published the first detailed maps of the seafloor, they showed undersea mountains rising above a flat abyssal plain and an underwater mountain range, called a mid-ocean ridge (see Modules 7.2 and 7.3), running the length of the Atlantic Ocean. The maps were based on echo soundings made by ships sailing across the Atlantic Ocean. Echo sounding is used to produce images of the seafloor by bouncing sounds off the seabed and working out how far the sound travels. As scientists discovered new features of the ocean floor (Figure 7.5.1), they developed theories to explain what they found.

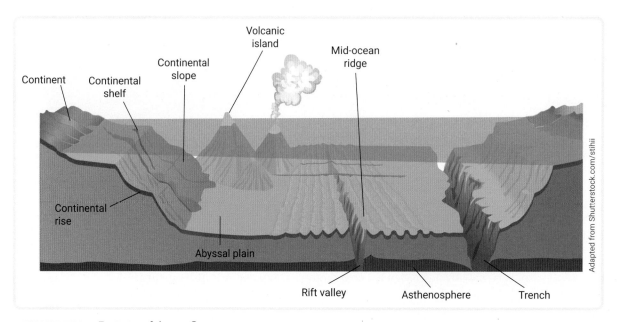

▲ **FIGURE 7.5.1** Features of the seafloor

seafloor spreading
the theory that new seafloor is created at mid-ocean ridges, spreads outwards and then descends into the mantle at trenches

In 1960, US researchers Robert Dietz and Harry Hess proposed a theory called **seafloor spreading**. They suggested that new seafloor is created at mid-ocean ridges, spreads outwards and then descends into the mantle at trenches. As the plates move apart over the asthenosphere, magma rises at the centre of the mid-ocean ridge and solidifies, creating new seafloor. As the plates continue to move outwards, more magma rises to form more seafloor (Figure 7.5.2).

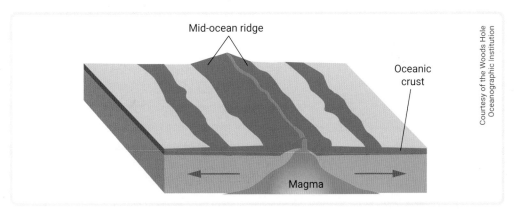

▲ **FIGURE 7.5.2** A mid-ocean ridge where new crust is formed. The stripes record changes in Earth's magnetic field (see Figure 7.5.4).

Evidence for seafloor spreading

Scientists soon found supporting evidence for seafloor spreading.

Heat flow and gravity

Surveys from ships across mid-ocean ridges noted that heat flow from the seafloor increased over the centre of the ridge. The central valley is a **rift**, created by the rocks being pulled apart. The increase in heat flow suggested that hot magma was close to the top of the mid-ocean ridge. Today we know that within the central valley of a mid-ocean ridge, basalt lava is squeezed onto the seafloor and hot water escapes from cracks in the crust called hydrothermal vents (Figure 7.5.3).

rift
a valley created by rocks being pulled apart

▲ **FIGURE 7.5.3** A hydrothermal vent

Magnetic patterns in ocean floor rocks

Just as magnetic patterns serve as evidence for the theory of plate tectonics (see Module 7.3), they also serve as evidence for seafloor spreading. Measurements of the magnetism of the seafloor show linear patterns parallel to and symmetrical about the mid-ocean ridge. The stripes reflect new seafloor being made, and continuously moving away from the ridge (Figure 7.5.4).

When a basalt lava or magma cools, magnetic minerals line up with Earth's magnetic field, and this shows the direction of the magnetic field. As the rock crystallises, the magnetite minerals preserve a record of the magnetic field at the time. The rock is then pushed outwards from the mid-ocean ridge and new seafloor is created.

Earth's magnetic field changes direction randomly every 200 000 to 300 000 years. When scientists map the normal magnetism and the **magnetic reversal** on a map, the different areas appear as symmetrical stripes parallel to the ridge.

Studies of the magnetism and age of lava flows on land allowed scientists to date parts of the seafloor where magnetic reversals occur. They found that ocean floor age increases as one moves further away from the ridge.

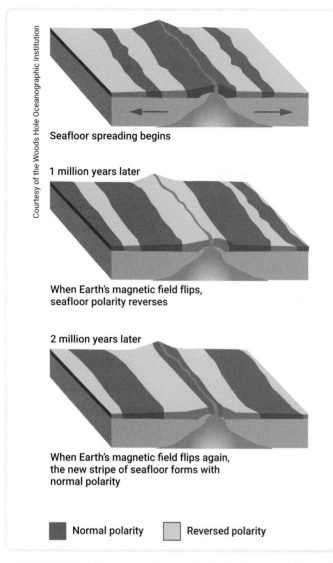

Seafloor spreading begins

1 million years later

When Earth's magnetic field flips, seafloor polarity reverses

2 million years later

When Earth's magnetic field flips again, the new stripe of seafloor forms with normal polarity

■ Normal polarity   ☐ Reversed polarity

▲ **FIGURE 7.5.4**   How magnetic seafloor 'stripes' are created

Dating of the seafloor

Magnetic reversals are not the only factor in studying the age of the seafloor. In the 1960s, a specially designed ship called the *Glomar Challenger* began drilling the seafloor (Figure 7.5.5). It drilled through sediments to the basaltic ocean crust below. The sediment contained fossil remains of single-celled organisms that once lived on the surface of the ocean (Figure 7.5.6). Because the ages of the fossils found directly above the basalt were known, they were used to determine the age of the ocean crust. The fossils confirmed that the rocks became older the further they were from the mid-ocean ridge.

▲ **FIGURE 7.5.5**   The *Glomar Challenger* drilled rock samples from the seafloor in the Atlantic Ocean, proving that the seafloor became older the further away it was from the mid-ocean ridge.

9780170463027

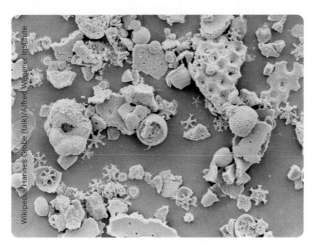

◀ **FIGURE 7.5.6** Fossils from deep ocean sediments. Each of these shells is less than 0.1 mm wide

magnetic reversal
a change in the direction of Earth's magnetic field

Wikipedia/Hannes Grobe (talk)/ Alfred Wegener Institute

☆ **ACTIVITY**

Age of the seafloor

Table 7.5.1 shows the distance from the Mid-Atlantic Ridge and oldest fossil age from five drill sites as collected by the 1968 deep-sea drilling program by the *Glomar Challenger*.

▼ **TABLE 7.5.1** Some samples collected by the *Glomar Challenger*, 1968

Location	Distance from Mid-Atlantic Ridge centre (km)	Approximate fossil age (millions of years)
14	727	40
16	191	11
17	643	33
18	506	26
19	990	49

What to do

1 Graph the age of the fossils against the distance to the Mid-Atlantic Ridge and draw a straight line that best fits through these points.

2 **Describe** the relationship between age and distance from the Mid-Atlantic Ridge axis.

What do you think?

1 How did this information support the idea of seafloor spreading?

7.5 LEARNING CHECK

1 Draw a labelled diagram to describe a mid-ocean ridge.
2 **Define** the theory of seafloor spreading.
3 **List** three pieces of evidence supporting the theory of seafloor spreading.
4 Why do maps of seafloor magnetism show stripes of normal magnetism and magnetic reversal?
5 How did deep-sea drilling of sediments support the theory that ocean floor varied in age?
6 **Explain** why ocean crust at the edge of an ocean is older than ocean crust formed at a mid-ocean ridge.

Video activities
What is a volcano?
What is an earthquake?

Extra science investigation
Modelling P and S waves

GET THINKING

Why do earthquakes and volcanoes occur at some places and not others? What causes them?

Plate boundaries, earthquakes and volcanoes

Each year, enormous amounts of energy are released around the world by earthquakes and volcanic eruptions. Most volcanoes and the origins of earthquakes are located along tectonic plate boundaries. Figure 7.3.3 (Module 7.3) shows the distribution of earthquakes and volcanoes around the world.

Volcanoes

A volcano is an opening in Earth's crust, through which molten rock reaches the surface (Figure 7.6.1). As we learned in Chapter 6, molten rock beneath Earth's surface is called magma, and magma that has erupted through Earth's surface is called lava. Magma contains dissolved gases, and when magma approaches Earth's surface, the gases can escape.

viscosity
a measure of a fluid's resistance to flow

The ability of magma to flow is affected by the amount of silicon dioxide present and the temperature of the magma. At divergent boundaries, magmas have a low **viscosity** and therefore flow readily. Gas escapes easily from the lava and explosions are rare. At convergent boundaries, magmas are more viscous. This is because the magma composition changes as some mineral crystals form and remove elements from the liquid. The remaining liquid increases in its silica dioxide content and its viscosity.

Shutterstock.com/WATHIT H

▲ **FIGURE 7.6.1**   Volcanoes such as these form at convergent boundaries.

The shapes of volcanoes and the types of eruptions they produce depend on the gas content and viscosity of the magma. At divergent boundaries, lava is hot and runny. It escapes from long fractures or cracks called **fissures** and flows easily. On land, these eruptions are called fissure eruptions (Figure 7.6.2). If the lava erupts from a central opening, called a **vent**, a shield volcano may form. Shield volcanoes are very large but have very shallow sloping sides. Mauna Kea, on Hawai'i, is an example of a shield volcano.

fissure
a long fracture or crack

vent
the central opening of a volcano

At convergent boundaries, volcanoes eject large amounts of ash through a central vent. The ash comprises small glassy fragments of volcanic glass created by expanding bubbles of trapped gas that break apart the rapidly cooling lava. The gases force the ash high into the sky.

The ash falls and creates a steep-sided volcano called a stratovolcano (Figure 7.6.3). A stratovolcano takes tens to hundreds of thousands of years to form. Its lavas are more viscous than those created at divergent boundaries.

▲ **FIGURE 7.6.2**   A fissure eruption

▲ **FIGURE 7.6.3**   A stratovolcano eruption

Earthquakes

An earthquake is a violent shaking of the ground caused by energy-carrying waves passing through Earth. Earthquakes can cause buildings and other structures to collapse. They can also cause landslides. At plate boundaries, enormous forces act on the rocks. Eventually rock breaks or rocks slide past each other on a fault. When they do, they release stored energy as an earthquake. Most earthquakes occur at plate boundaries, but earthquakes regularly occur away from plate edges. The cause is the same as at a plate boundary, but there is usually less energy released.

focus
the place where earthquake energy is released

epicentre
the point on Earth's surface directly above an earthquake focus

The place where earthquake energy is released is called the **focus** (plural: foci) of the earthquake. The point on the Earth's surface directly above the focus is called the **epicentre**. In subduction zones, the earthquake focus may be as deep as 700 kilometres below the surface.

As the energy moves away from the focus, it travels as different types of waves. Earthquake waves that travel deep below the surface are called body waves, and waves that travel close to the surface are called surface waves (Figure 7.6.4). Surface waves cause the greatest shaking in an earthquake. Fortunately, they become less energetic as they move further away from the focus.

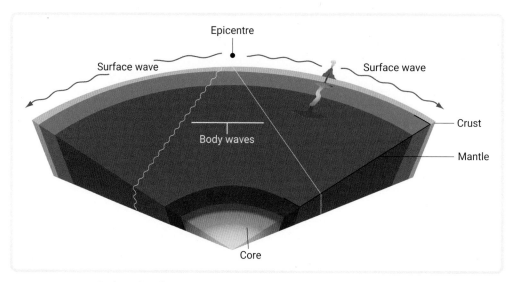

▲ **FIGURE 7.6.4**   Body and surface waves

Earthquakes are measured by the energy or by the effects they produce. The energy is measured as an earthquake magnitude. The effects are described as intensities. Small earthquakes (magnitude 0–2) are very common but huge ones (magnitude 8 or higher) are very rare (Table 7.6.1). The magnitude scale does not increase in a regular way. Each increase of 1 in the scale is an increase of 10 times the energy. This means a magnitude 3 earthquake has 10 times more energy than a magnitude 2 earthquake and a magnitude 4 earthquake has 100 times more energy than a magnitude 2 earthquake. Each year there are about 100 earthquakes of magnitude 3 or higher in Australia.

▼ **TABLE 7.6.1**   Global earthquake magnitudes and frequencies

Earthquake magnitude	How many occur in a year?	Earthquake effects
Less than 2.0	Several million	Rarely seen or felt by people
2.0–2.9	More than a million	A few people see and feel it
3.0–3.9	More than 100 000	Many people see and feel it; ceiling lights swing
4.0–4.9	More than 10 000	Most people see and feel it; walls crack
5.0–5.9	More than 1000	Damages buildings near epicentre; furniture moves
6.0–6.9	More than 100	Causes great damage around the epicentre
7.0–7.9	10–20	Damages most buildings
8.0–8.9	1	Causes major damage to buildings and other structures
9	About one every 10 years	Causes most buildings to collapse and destroys bridges and roads

9780170463027

Tsunamis

Earthquakes are the major cause of large powerful ocean waves called **tsunamis**. Most tsunamis are generated by the sudden movement of the seafloor at convergent boundaries. The sudden movement causes the water above the seafloor to move up or down. As the water flows back, it generates a series of waves that rapidly move outwards. Tsunamis can also be created by undersea landslides, volcanic explosions or meteorite impacts.

tsunami
a series of large ocean waves created by an undersea earthquake or volcanic eruption

A tsunami causes damage when it reaches the shore. As it approaches the shore, the waves slow and increase in height (Figure 7.6.5). When the waves reach the shore, they surge inland, destroying buildings and carrying away anything that cannot resist the surge of water.

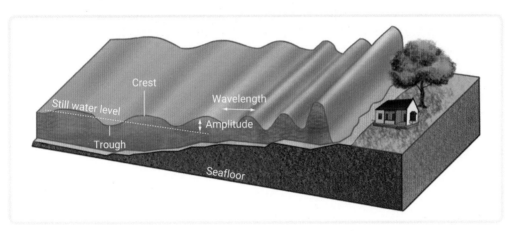

▲ **FIGURE 7.6.5** As a tsunami enters shallow water, the waves slow down but their height (amplitude) increases.

In March 2011, an area of seafloor 300 km long and 150 km wide was thrust 10 metres upwards, causing a magnitude 9.0 earthquake. This created a tsunami that rapidly moved away from the epicentre. The Japanese city of Sendai was only 180 kilometres from the epicentre. The tsunami was 11 metres high when it reached the coast and it travelled up to 10 kilometres inland, causing enormous damage and loss of life. More than 19 000 people died.

7.6 LEARNING CHECK

1 **Define** the following terms.
  a Earthquake
  b Volcano
  c Tsunami
2 How is lava different from magma?
3 **Name** the plate boundary type where stratovolcanoes are found.
4 **Describe** where earthquakes occur and explain why they occur there.
5 **Explain** why damage to coastal communities can occur due to a tsunami and the earthquake that caused it.

7.7 First Nations Australians' records of geological events in Australia

FIRST NATIONS SCIENCE CONTEXTS

IN THIS MODULE, YOU WILL:

✓ explore how the cultural narratives of First Nations Australians provide records of geological events that occurred in Australia many thousands of years ago.

Cultural narratives of volcanic eruptions

Although Australia has no active volcanoes and damaging earthquakes don't happen very often, this wasn't always the case. Tens of thousands of years ago, First Nations Australians experienced active volcanoes and major earthquakes that helped to shape Australia.

With more than 60 000 years of recorded habitation, First Nations Australians have one of the longest continuing cultural histories connected with a single geographical region in the world. Their knowledge has been passed from generation to generation through oral traditions and it provides details of geological events that can be dated back tens of thousands of years.

▲ **FIGURE 7.7.1** Budj Bim is an extinct volcano in south-west Victoria.

The Gunditjmara Peoples of south-west Victoria are the traditional owners of the Budj Bim World Heritage-listed area, which includes the Budj Bim volcano (Figure 7.7.1). The Gunditjmara Peoples have a cultural narrative that explains the formation of their Country. It describes four giants who gave life and laws to the land. During this period, one ancestral being, Budj Bim, emerged and revealed himself to the surrounding landscape during a volcanic eruption. Lava produced by the volcano has been dated at around 37 000 years old, so it is thought that the narrative could be one of the oldest ever told. The description in the narrative may also demonstrate deductive reasoning about the local geology by the Gunditjmara Peoples.

The Kinrara crater is in north Queensland. Cultural narratives of the Gugu Badhun Peoples tell of a pit forming in the plains, dust filling the air and a river of fire emerging from the ground. The lava flows, which were up to 55 kilometres long, are still clearly visible. The volcanic rocks formed from the eruption of Kinrara have been dated at around 7000 years old, making this one of Australia's youngest eruptions. Some people have suggested these dates potentially mean the story of the eruption has been passed down through more than 230 generations, a period longer than the earliest written records.

Recent volcanic activity in far north Queensland has been dated to more than 10 000 years ago. Cultural narratives describing the volcanic eruptions that formed Eacham (Figure 7.7.2), Barrine and Euramo crater lakes in far north Queensland are well documented.

Alamy Stock Photo/UtCon Collection

9780170463027

The Ngadjon-Jii Peoples' cultural narrative explains how two men broke important Ngadjon-Jii laws. This angered the rainbow serpent, an important ancestral being. The narrative explains how the rainbow serpent caused an earthquake and describes the dynamic nature of these events, such as loud noises, violent shaking and the cracking of the earth. The Ngadjon-Jii Peoples' narratives contain not only geological knowledge of volcanic events in the area, but also ecological knowledge. Although the region is now tropical rainforest, the narratives describe it as being covered by eucalypt scrub. Analysis of fossil pollen found in the silt of the crater lakes shows the current rainforest to be approximately 7600 years old. This analysis reaffirms the accuracy and reliability of this narrative, which has impressively remained consistent over more than 200 generations. For many, this illustrates the power of cultural narratives as a way of accurately communicating information across many generations.

▲ **FIGURE 7.7.2** Lake Eacham in far north Queensland was once the crater of a volcano.

☆ **ACTIVITY 1**

1 For one of the cultural narratives above:
   a relate the events in the narrative to events associated with a volcanic eruption.
   b describe what geological evidence would be present today to confirm that a volcanic eruption occurred.

Narratives of earthquakes and tsunamis

Knowledge of earthquake activity in the Newcastle region is contained in the men's business oral narratives of the Awabakal Peoples.

The narratives of the Gundungurra Peoples of south-eastern New South Wales and the Kambure Peoples of the Kimberley tell of tsunami events that inundated coastal regions.

In New South Wales, the Barkandji Peoples have a narrative that tells of a great ball of fire falling from the sky, causing tremendous flooding. The narrative then says that the people only survived by running up into the hills. On the eastern

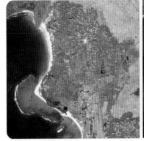

▲ **FIGURE 7.7.3** Aceh, Indonesia, before (left) and after (right) the huge 2004 tsunami that claimed the lives of 220 000 people.

seaboard of Australia, scientists have found sedimentary rock layers in the area, dated at around 1500 CE, which show a disturbance that corroborates the oral records of a tsunami occurring in this region. Maori narratives tell of a similar event during a similar timeframe. In 2003, geologists found a 20-kilometre diameter submarine structure in the Tasman Sea south of New Zealand, believed to be an impact crater, with an estimated impact date of approximately 1443 CE and enough energy to have produced a tsunami. This find supports the cultural narratives of a tsunami around that time.

☆ **ACTIVITY 2**

1 Relate the events in the Barkandji Peoples' cultural narrative to the events associated with a tsunami.

BY THE END OF THIS MODULE, YOU WILL BE ABLE TO:

✓ describe major developments in the study of the seafloor and plate tectonics.

Video activity
Mapping the secrets
of the ocean floor

Our understanding of plate tectonics and how the seafloor forms and changes has evolved over a very long time. Below is a summary of major discoveries in the development of plate tectonics.

▼ TABLE 7.8.1 A brief history of seafloor discovery and plate tectonics

1872–76	The British Challenger expedition gathered rocks from the seafloor and made depth measurements with weights on ropes, leading to the discovery of ocean trenches. Until this time, people thought the seafloor was flat, ancient and unchanging.
1930s and 1940s	Scientists used echo sounders to discover many young undersea mountains.
1947	Maurice Ewing discovered that the ocean crust is mainly basalts, rather than granites as many people expected.
1950	US scientists discovered that the seafloor has a thin sediment layer and a uniform ocean crust structure.
	Scientists detected magnetic variations in the ocean seafloor.
1953	Marie Tharp and Bruce Heezen published a map showing an underwater mountain range: the Mid-Atlantic Ridge.
1955	Geologists discovered magnetic striping in the ocean floor, parallel to and symmetrical about mid-ocean ridges.
1956	Scientists found a deep rift valley in the Mid-Atlantic Ridge.
1960	Harry Hess and Robert Dietz proposed the seafloor spreading hypothesis.
1963	Frederick Vine, Drummond Matthews and Lawrence Morley tested the seafloor spreading theory using magnetic data. Their work supported the theory and provided a way to measure the rate of seafloor spreading.
1964	George Plafker studied a large earthquake in Alaska and concluded that the earthquake was caused by the Pacific crust being forced under, or subducted, beneath Alaska.
1965	Canadian J. Tuzo Wilson described and explained transform faults.
1967	Scientists explored the idea of tectonic plates moving relative to each other.
1968	The *Glomar Challenger* drilled cores adjacent to the Mid-Atlantic Ridge, confirming that ocean crust is young and forms at the mid-ocean ridge.
1995	GEOSAT satellite radar data allowed researchers to map the world's entire ocean floor.

☆ ACTIVITY 1

Plate tectonics timeline

What to do

1 **Identify** when the following discoveries or concepts occurred using the summary in Table 7.8.1.

   a Ocean crust is ancient and unchanging.

   b Ocean crust is young.

   c Ocean crust is created at mid-ocean ridges.

   d Ocean crust is not the same as continental crust.

   e Trenches

   f Subduction zones

g Mid-ocean ridges   **i** Magnetic striping

h Transform faults   **j** Seafloor spreading hypothesis

2 Use the information from your answer to question 1 to create a timeline from 1850 to 2000 of these discoveries and concepts.

What do you think?

1 **Identify** three methods of gathering data that were important in developing the plate tectonics theory.

2 What does your timeline show about the rate of discovery up until 1950?

3 **Summarise** how our understanding of the ocean floor is different from what people thought in the 1800s.

Calculating the rate of plate spreading

☆ ACTIVITY 2

Using knowledge of when magnetic field reversals happen, scientists can identify the age of the seafloor near a mid-ocean ridge. In this exercise you will use second-hand data to work out the speed at which an ocean is opening.

What to do

1 Examine Table 7.8.2.

▼ TABLE 7.8.2   Age of seafloor at different distances from a mid-ocean ridge

North of the ridge		South of the ridge	
Distance from mid-ocean ridge (km)	Age of the seafloor (millions of years)	Distance from mid-ocean ridge (km)	Age of the seafloor (millions of years)
0	0	0	0
94	2.6	94	2.6
528	19	470	19
605	23	504	23
1206	50	1180	50
1547	66	1560	66

2 Use graph paper to graph the data north of the mid-ocean ridge. Plot distance on the vertical axis and age on the horizontal axis. Remember to label your axes correctly.

3 Draw a line of best fit through the points.

4 To calculate the average speed that the seafloor is moving, calculate the slope of the line of best fit. Your answer will be in kilometres per million years or millimetres per year.

5 Repeat steps 1–3 for the data south of the mid-ocean ridge.

What do you think?

1 Are the average speeds of the plates north and south of the mid-ocean ridge similar?

2 Is the ocean spreading at a constant rate? How can you tell?

3 If the ocean is between two continents, how much further apart do the two continents move each year?

Researching and presenting scientific data

SCIENCE INVESTIGATIONS 7.9

SCIENCE SKILLS IN FOCUS

IN THIS MODULE, YOU WILL FOCUS ON LEARNING AND IMPROVING THESE SKILLS:

▶ represent collected data using different media to describe patterns and trends

▶ identify valid scientific conclusions based on collected data

▶ make and assess predictions based on evidence and scientific knowledge.

Video
Science skills in a minute: Presenting data in different forms

Science skills resource
Science skills in practice: Data in different forms

Maps are a great way to organise and process large amounts of data. Maps present information as a picture rather than a series of numbers.

▶ **Five characteristics of a good map are:**

• **title** – make your title descriptive, so that the purpose of your map is clear

• **scale** – indicate a scale so the reader knows the distances represented on the map

• **legend** – a legend is a key to symbols on a map. A key is a list of symbols and names of what the symbols represent. Use clear, simple symbols and choose colours that can easily be differentiated apart

• **compass** – a compass shows which way is North, so that a reader can orient the map easily

• **lines of latitude and longitude** – these lines can help a reader work out where on Earth your map is located.

INVESTIGATION 1: USING MAPS TO SUMMARISE INFORMATION AND SHOW PATTERNS

AIM

To create a map showing the location of convergent boundaries in the Pacific and East Asia

YOU NEED

☑ a world map with tectonic boundaries marked on it, covering the Pacific and East Asia

☑ coloured pencils

WHAT TO DO

Convergent boundaries involving oceanic crust have:

• trenches

• fold mountains

• zones of shallow to deep earthquakes

• stratovolcanoes erupting ash.

1   On the world map, identify and mark with a blue pencil the location of ocean trenches.

2   Create a key in the bottom centre of the map. Add a blue line and the word 'Trenches' to your key.

3   Study the earthquake distribution in Figure 7.3.3 (page 190). Remember that convergent boundaries are marked by wide zones of descending (shallow to deep) earthquake foci. Use at least one other information source to check where such zones of shallow to deep earthquakes occur.

4   Use a different coloured pencil and shade in zones of descending earthquakes. Add your shading and a name to the key.

5   Search the Internet for the National Centers for Environmental Information Natural Hazards Viewer to locate volcanoes around the Pacific, in the Philippines and in Indonesia.

6   Use a new pencil colour and mark on your map the zones where the volcanoes are found. Do not try to add every volcano. Add your symbol to the key.

7 Use Figure 7.2.2 (page 185) to add arrows on your map showing the direction of movement for the Australian, Nazca and Pacific plates.

9 Add a title to your map.

WHAT DO YOU THINK?

1 How closely related are the earthquakes, volcanoes and trenches?

2 Do the directions of plate motion match the location of the convergent boundaries?

3 Why is a key important in making your map understandable?

4 Explain why a map like this is more useful than a table of data.

INVESTIGATION 2: VOLCANIC ERUPTIONS

AIM

To find, process and present information on a famous volcanic eruption

YOU NEED

☑ access to the Internet and library resources

WHAT TO DO

1 Select a volcanic eruption from the list below.
- Mount Vesuvius, 79 CE
- Mount Agung, 1963
- Mount St Helens, 1980
- Mount Pinatubo, 1991
- Laki, 2011
- Kilauea, 2021
- La Palm, 2021

2 Research information on the eruption to describe:
  a where it occurred.
  b the effect the event had on surrounding areas.

3 Select a format (poster, slide show etc.) in which to present your research.

4 Create your presentation so that you can share it with other students.

INVESTIGATION 3: TSUNAMIS AND EARTHQUAKES

AIM

To determine the relationship between convergent plate boundaries and tsunamis

WHAT YOU NEED

☑ access to the Internet and library resources

WHAT TO DO

1 Research the tsunami that occurred on 26 December 2004. What caused the tsunami? What countries did the tsunami affect? What damage did the tsunami cause?

2 Research the 2011 Tōhoku earthquake and tsunami. Where did the earthquake and tsunami originate? What damage did the tsunami cause?

3 Volcanic explosions can also cause tsunamis. In 1883, the explosion of Krakatoa in Indonesia created a tsunami 30 metres high. Investigate the effects of the tsunami. What was the greatest distance at which the tsunami was detected?

WHAT DO YOU THINK?

1 Were the sources of these tsunamis near a plate boundary? If so, what type was it?

2 Which tsunami caused the greatest loss of life and damage?

3 Which of the causes of a tsunami is most likely to occur again? Why do you think so?

CONCLUSION

Predict the type of plate boundary where tsunamis are most likely to be created.

> **Warning**
>
> Use tongs and take care when handling hot items. Don't eat the food used in the investigation.

INVESTIGATION 4: CHEESY PLATE TECTONICS

AIM

To model the layers of Earth and demonstrate the movement of tectonic plates

YOU NEED

- ☑ 2 tortilla wraps
- ☑ 2 slices of cheese
- ☑ large piece of aluminium foil
- ☑ hot plate
- ☑ metal tray
- ☑ tongs
- ☑ heatproof mat or ceramic tile

WHAT TO DO

1 Place the aluminium foil onto the metal tray.

2 Place two slices of cheese on the piece of aluminium foil, approximately half a tortilla wrap's distance apart.

3 Place a tortilla wrap on top of each of the slices of cheese, ensuring the edges of the tortillas are just touching.

4 Place the metal tray onto the hot plate and begin to slowly heat it.

5 Heat until the cheese is completely melted, then use the tongs to remove the metal tray from the hot plate and place on a heatproof mat or ceramic tile.

6 Using the tongs in one hand to keep the tray steady (and being careful not to touch the tray), place the other hand on top of one of the tortilla wraps and attempt to slide it sideways into/on top of the other tortilla wrap. Take note of what happens and record the results.

7 Now place one hand on top of each of the tortillas. Slide them slowly apart as far away as possible from each other. Take note of what happens to the cheese and record the results.

RESULTS

Copy and complete a table like the one below.

Activity	Observation
Sliding one tortilla into the other	
Sliding tortillas apart	

1 Which layers of Earth do you think the tortilla and the cheese represent, respectively?

2 What does our activity of sliding one tortilla into the other represent on Earth?

3 Where on Earth does this kind of event occur?

4 What happened to the cheese when you slid the tortillas away from each other? What do you think this represents on Earth?

CONCLUSION

Write a conclusion about how your tectonic plate experiment relates to real-life phenomena that occur on Earth. Use your observations to support your arguments.

REMEMBERING

1 What name is given to each of the following?
   a The living parts of Earth
   b All the water in the world
   c The largest rocky layer of Earth
   d The part of Earth composed of solid iron and nickel
   e Gases held within Earth's vicinity by gravity
   f The part of Earth composed of liquid metal

2 List three differences between continental crust and oceanic crust.

3 Name the type of plate boundary where:
   a plates slide past each other.
   b steep sided, ash-erupting volcanoes occur.
   c rift valleys form at the centre of mountain chains.
   d low-viscosity lavas erupt.
   e deep earthquakes and trenches occur.

4 State the theory of continental drift.

5 State the theory of plate tectonics.

6 Describe the cause of a tsunami.

UNDERSTANDING

7 Draw a labelled diagram to show how convection in Earth's mantle occurs.

8 Why is oceanic lithosphere less than 180 million years old when oceans have existed for billions of years?

9 The image below shows the interactions of some tectonic plates. Use this information to answer the following questions.
   a Which plate contains the oldest lithosphere?
   b On which plate would you find the epicentres of deep earthquakes?
   c At which boundary do shallow earthquakes occur, but no volcanic activity?

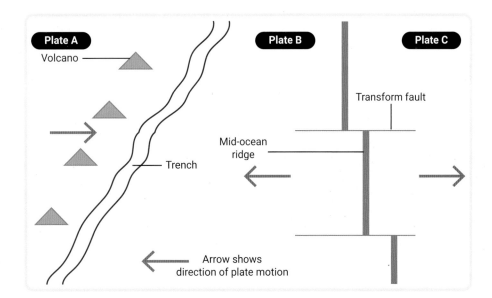

10 Use a table to compare the locations, effects and causes of slab pull and ridge push.

11 How did the presence of rocks resulting from glaciers lead Alfred Wegener to conclude that Pangaea once existed?

12 List the features you would look for to identify a convergent boundary.

13 Explain why an earthquake could occur at a fault that was not at a plate boundary.

14 'Tectonics' comes from a Greek word meaning *to build or construct*. How do tectonic plates construct features on Earth's surface?

APPLYING

15 Create a diagram of a divergent boundary and show how stripes of normal and reversed magnetism would be arranged around the boundary.

16 Why is a tsunami likely to cause more damage approaching a coast with a shallow beach than a coast with deep water?

17 What are three pieces of evidence to show that two continents were once joined together?

18 Explain whether it would be possible for plate tectonics to occur on a planet with a cold mantle.

19 Write a statement relating the viscosity of lava and the type of volcanic eruption it produces.

20 As ocean crust cools, it thickens and becomes denser. Use this information and your knowledge of the asthenosphere to explain why the ocean becomes deeper as ocean crust moves further from the mid-ocean ridge.

21 Dr Duane Hamacher from the UNSW Indigenous Astronomy Group uncovered evidence linking First Nations Australians cultural narratives about meteor events with impact craters dating back 4700 years. **Suggest** why he made the statement: 'Aboriginal stories could lead us to places where natural disasters occurred.'

EVALUATING

22 Why didn't many scientists believe in continental drift when Alfred Wegener had so much evidence to support his theory?

23 If convection occurs in Earth's mantle, explain why it might also occur in the outer core.

24 Why do earthquakes originate in the subducting plate at a subduction zone?

25 The Himalayas and the Andes mountains are both the result of plate convergence. Why are there no active volcanoes in the Himalayas?

26 As the magma chamber below a volcano cools, the magma becomes cooler and richer in silica and dissolved gases. Predict how these changes would affect the type of volcanic eruptions the volcano produces in the future.

CREATING

27 Create a concept map to show the relationships between plate boundaries, earthquakes, volcanic activity and geographic features such as mountains and trenches.

28 Volcanoes that erupt basaltic lava and rifts that are caused by tension forces are common on other planets. However, scientists have not yet found evidence of plate tectonics on other planets. Outline the types of evidence for plate boundaries that scientists should look for if they were to discover plate tectonics on other planets.

BIG SCIENCE CHALLENGE PROJECT #7

1 **Connect what you have learned**

In this chapter you have learned how plate boundaries shape the surface of Earth and some of the structures that tectonic forces build. Iceland is an island built by a hotspot on top of a divergent boundary. Review the chapter and make a mind map of the features (volcanic activity, earthquakes, geography etc.) you might expect to find in Iceland.

2 **Check your thinking**

Explain, in terms of plate tectonics, what is happening to form the rift valley in the photograph. What sort of volcanic activity has created the cliffs on the far side of the valley? What types of earthquake activity would you expect to find in Iceland?

Shutterstock.com/Vaclav Sebek

3 **Make an action plan**

Make a plan to compare Iceland with Australia. Conduct some research about the geology of Iceland. What is the nature of volcanoes on Iceland? What sorts of features are found on Iceland that you might expect to find at a divergent boundary?

Research the geology of Australia. How common are active volcanoes in Australia? Does Australia have mountains? Has it had mountains in the past? Why is Australia such a dry, flat country?

4 **Communicate**

Use your knowledge and understanding to create a table contrasting the tectonic features of Iceland with the tectonic features of Australia. Your table should have three columns with headings: 'Characteristics being compared', 'Iceland' and 'Australia'.

8 Energy

8.1 **Kinetic energy** (p. 214)

A mass in motion has kinetic energy.

8.2 **Potential energy** (p. 218)

Potential energy is stored energy ready to do work.

8.3 **Energy transfer** (p. 222)

Energy transfer is the movement of a single type of energy from one place to another or from one body to another.

8.4 **Energy transformation** (p. 226)

Energy transformation is changing energy from one type into another.

8.5 FIRST NATIONS SCIENCE CONTEXTS: **First Nations Australians' traditional fire-making techniques** (p. 230)

First Nations Australians used their knowledge of energy transfer and friction to develop methods for making fire.

8.6 SCIENCE AS A HUMAN ENDEAVOUR: **Generating electricity in remote communities** (p. 233)

Remote communities rely on small electrical generation systems such as RAPS to meet their electricity needs.

8.7 SCIENCE INVESTIGATIONS: **Modelling energy transfers and transformations with flow charts** (p. 234)

Modelling energy transfers and transformations

9780170463027

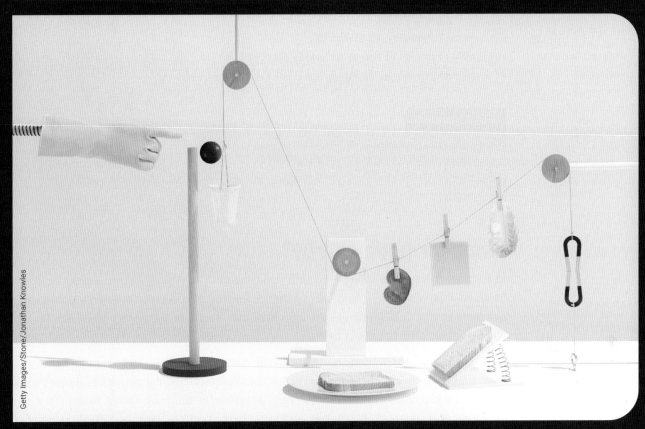

Getty Images/Stone/Jonathan Knowles

▲ **FIGURE 8.0.1** An example of a Rube Goldberg machine

A Rube Goldberg machine is a machine that uses many energy transfers and transformations to complete a simple task in a complicated way. Energy is a basic idea in science, but what is energy? How is energy defined? What is an energy transfer or transformation? Why are they important?

▶ **Think about your lifestyle. What types of energy do you use every day?**

#8   SCIENCE CHALLENGE ACCEPTED!

At the end of this chapter, you can complete Big Science Challenge Project #8. You can use the information you learn in this chapter to complete the project.

Assessments
- Prior knowledge quiz
- Chapter review questions
- End-of-chapter test
- Portfolio assessment task: Project

Videos
- Science skills in a minute: Organising data into charts **(8.7)**
- Video activities: Forms of energy **(8.2)**; Rube Goldberg machine **(8.3)**; Energy transformation **(8.4)**; Clean energy in remote communities **(8.6)**

Science skills resources
- Science skills in practice: Energy transfers and transformations **(8.7)**
- Extra science investigations: Transferring and transforming energy **(8.4)**; Transforming gravitational energy **(8.4)**

Interactive resources
- Simulations: Kinetic and potential energy **(8.2)**; Energy transfers **(8.3)**
- Drag and drop: Types of kinetic energy **(8.1)**; Renewable or non-renewable? **(8.4)**
- Match: Types of potential energy **(8.2)**
- Label: Energy transfer diagrams **(8.3)**

⁂ Nelson MindTap

To access these resources and many more, visit:
cengage.com.au/nelsonmindtap

Interactive resource
Drag and drop:
Types of kinetic
energy

What is kinetic energy?

In science, **energy** is defined as the ability to do **work**. The type of energy is described
by where it comes from or how we experience the energy. One type of energy is known
as **kinetic energy**. It is defined as the energy of movement, or the energy possessed by
objects or particles that are in motion.

Energy is measured using the metric unit called the **joule (J)**. The joule is named
after scientist James Prescott Joule (1818–89), who completed many investigations
about energy. The joule is a very small unit, so energy is more commonly measured in
kilojoules (kJ), a unit of 1000 joules. You may also be familiar with the imperial unit of
calorie (cal) that is still often referred to in food energy values.

energy
the ability to do work, such
as moving or cooking, or
the potential to do work,
such as the chemical
potential energy stored
in food

work
energy that is being
transferred or transformed

kinetic energy
the energy possessed by
an object or particle that
is in motion

joule (J)
the metric unit of
measurement for energy

calorie (cal)
the imperial unit of
measurement for energy

▲ **FIGURE 8.1.1**   Skateboarding is an example of kinetic energy in action.

Types of kinetic energy

There are several types of energy that are examples of kinetic energy. All of these share the common characteristic of movement of an object or particles. Some of these types of energy may be obvious to you, whereas others may not. Some motion we may be able to see, such as a moving truck or car, whereas other motion is not visible. Can you think of forms of energy where we cannot directly sense particles that are in motion?

sound energy
energy transferred as a wave through or by vibrations

Sound energy

Sound energy is a form of kinetic energy where the energy is transferred as a wave. Sound energy from an explosion can break windows or other glass (Figure 8.1.2). We use sound energy to communicate when we talk to one another. Consider what else you know about sound.

▲ **FIGURE 8.1.2**   Sound energy can break glass.

☆ ACTIVITY 1

Observing sound energy

You need
- tuning fork
- Petri dish
- water

What to do

1 Strike the tuning fork on a hard surface and place the tuning fork close to your ear. What can you sense?

2 Touch the tuning fork on your earlobe. What do you sense now?

3 Touch the prongs of the tuning fork to the surface of a Petri dish of water. What do you observe?

What do you think?

Explain why you can see the sound energy in the water.

Heat and thermal energy

When hot objects are brought close to colder objects, heat is transferred between them. You have probably experienced walking on the ground or sand in bare feet during summer (Figure 8.1.3). Why does the ground feel hot? This experience shows you that hot objects transfer heat to colder objects.

▲ **FIGURE 8.1.3**   Hot sand transfers heat to your (colder) skin.

Observing thermal energy

You need

- metal rod or ruler
- adhesive putty (Blu Tack)
- drinking cup
- hot water

> ⚠ **Warning**
> Hot water can scald you. Take care when working with hot water in the lab.

What to do

1 Attach a piece of adhesive putty about halfway along the metal rod. Feel the temperature of the metal.

2 Place one end of the metal rod into a cup of hot, but not boiling, water. Wait a few minutes and describe what happens.

3 Carefully take the metal rod out of the water. Describe its temperature now compared to earlier.

What do you think?

Explain why there has been a change in temperature.

thermal energy
energy contained within an object by its vibrating particles, which determines an object's temperature

heat
thermal energy of an object that can be transferred to another object

electrical energy
energy carried by charged particles in electric circuits

In Activity 2, at first the metal rod probably felt cooler than your hand. When you placed the metal into the hot water, the water transferred some of its **thermal energy** to the rod. The adhesive putty gained energy from the metal rod, heated up and may have fallen off the metal rod.

As you learned in Year 7, the particle theory states that all particles are moving to some degree. As a substance is heated, the movement of the particles increases and the temperature of the substance increases. When the substance cools, the particles slow down, and the substance's temperature will slowly decrease. When thermal energy is transferred from one object to another, this is correctly referred to as **heat**. Heat is the energy transferred from one object to another because of a difference in temperature.

Electrical energy

One of the most common forms of kinetic energy that we all use every day is **electrical energy**. We usually refer to this as electricity. Appliances and devices are plugged into power points to receive electricity to make them work (Figure 8.1.4). Electrical energy is transferred by charged particles moving through a conducting circuit. Some devices may not need to be connected to a power point to work. These devices use batteries as the source of electrical energy.

▲ **FIGURE 8.1.4** We commonly use electrical energy in our everyday lives.

Light energy

Light energy is another form of kinetic energy. We can see light directly or when it is reflected off surfaces or **particulates** in the air, such as dust (Figure 8.1.5). Light behaves quite differently from other types of kinetic energy because it is a wave. You will study light and sound waves in more detail in Year 9.

light energy
energy transferred as a form of electromagnetic radiation

particulates
very small pieces of substances suspended in air or in water

▲ FIGURE 8.1.5   We can see light when it is reflected off particulates in the air.

8.1 LEARNING CHECK

1 **List** three examples of objects that possess kinetic energy.

2 a **Describe** two types of energy that are classified as kinetic energy.

  b **Explain** why your examples are types of kinetic energy.

3 **State** what makes light different from other types of kinetic energy.

4 **Construct** a table that summarises the type of kinetic energy and the source of the movement observed. Use the headings 'Type of kinetic energy' and 'Source of movement'.

✓ describe different types of energy

✓ classify examples of energy as potential energy

✓ explain why examples of energy are classified as potential energy.

Video activity
Forms of energy

**Interactive
resources**
Simulation: Kinetic
and potential energy
Match: Types of
potential energy

GET THINKING

After completing this module, you will be able to compare examples of energy and
determine whether they are types of kinetic or potential energy. As you work through
this module, make a mental note of the differences between types of kinetic energy and
potential energy.

potential energy
energy that is stored,
ready to be changed to
another type of energy

What is potential energy?

Potential energy is energy stored within an object or material. Energy that is stored
in some way has the potential to make something happen or to do work. Much of the
energy around us is in a type of potential energy.

Types of potential energy

There are several types of energy that are examples of potential energy. All involve
stored energy that has the potential to do work. Can you think of types of energy that
are stored?

Gravitational potential energy

**gravitational potential
energy**
energy associated with
the position of an object
above a reference height

There are examples in the news that describe objects falling as a result of some event. In
mountainous regions of the world, avalanches occur regularly. Avalanches occur because
material, such as ice, has **gravitational potential energy**. Gravitational potential energy
(usually just referred to as *gravitational energy*) is the potential energy associated with
how high an object is above some reference point. As an object is lifted higher above
its starting height, its gravitational energy
increases. As the object falls, gravitational
energy is converted to kinetic energy. When
the object hits the ground, it has no remaining
gravitational energy. People who skydive use
gravitational energy to fall when they leave the
plane. Skydivers convert gravitational energy
into kinetic energy as they fall towards the
earth. When unstable topsoil on a mountain
or hill becomes saturated with water, the extra
weight can cause it to slide away from the
ground or rocks beneath it in a landslide or
mudslide (Figure 8.2.1). As rocks and mud
fall, there may be some energy transfer as
this material collides with other material and
transfers some of its kinetic energy.

▲ **FIGURE 8.2.1** A landslide caused by heavy rain and storms
completely blocked a local road in the Bilgola Plateau in Sydney,
March 2022.

Chemical potential energy

There are other types of potential energy. For example, where do you get the energy to do all the things you do every day? This energy comes from food (Figure 8.2.2). The food we eat contains chemicals that our body converts into energy through chemical reactions. This energy is called **chemical potential energy** (usually referred to as *chemical energy*). You learned about energy in chemical reactions in Chapter 5. The process of supplying energy to your body is called cellular respiration, and occurs in every cell of living things. You learned about this process in Chapter 2. The energy supplied enables your body to carry out its natural processes, such as thinking, breathing and moving the blood around your body, as well as doing physical work, such as walking and talking.

On the labels of most packaged foods is a table that lists the main groups of chemicals contained in the food, and how much chemical energy is released from consuming a measured quantity of that food (Figure 8.2.3).

▲ **FIGURE 8.2.2** This meal contains chemical energy that enables you to do work.

chemical potential energy
energy stored in the chemical bonds of a substance and released when the substance reacts

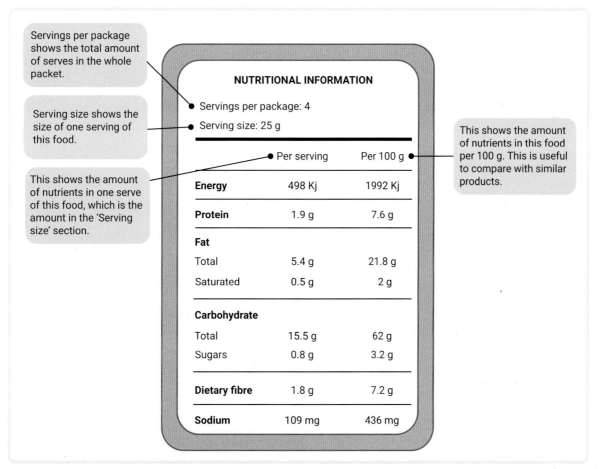

Servings per package shows the total amount of serves in the whole packet.

Serving size shows the size of one serving of this food.

This shows the amount of nutrients in one serve of this food, which is the amount in the 'Serving size' section.

This shows the amount of nutrients in this food per 100 g. This is useful to compare with similar products.

NUTRITIONAL INFORMATION

Servings per package: 4
Serving size: 25 g

	Per serving	Per 100 g
Energy	498 Kj	1992 Kj
Protein	1.9 g	7.6 g
Fat		
Total	5.4 g	21.8 g
Saturated	0.5 g	2 g
Carbohydrate		
Total	15.5 g	62 g
Sugars	0.8 g	3.2 g
Dietary fibre	1.8 g	7.2 g
Sodium	109 mg	436 mg

▲ **FIGURE 8.2.3** A typical food label used in Australia

Examining food labels

What to do

1 Draw a table with three columns with headings 'Food group', 'Energy value' and 'Quantity consumed'.

2 Include energy value information from the labels of food and drinks that you consume throughout the day. Make a note of how much of each type you consume.

3 Complete the table for two consecutive days and compare what you have recorded.

What do you think?

1 Which groups of foods have the highest energy values?

2 Which groups of foods have the lowest energy values? You may be surprised by the chemical energy contained in fast foods.

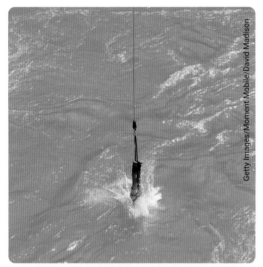

▲ **FIGURE 8.2.4**   The bungee cord stores energy as elastic potential energy. It has the potential to release the energy and pull the person at the end of the cord back upwards.

elastic potential energy
energy stored in a spring or elastic material

Elastic potential energy

When you squash or stretch a spring, the energy you use to squash or stretch it is stored in the spring as **elastic potential energy** (usually referred to as *elastic energy*). This energy does work when the spring is released. When you stretch a material, such as an elastic band, the energy you used to stretch it is stored in the elastic. It is possible to release this energy later. A bungee cord is made of a spring-like elastic material. The cord stretches and stores the energy as potential energy (Figure 8.2.4).

Elastic energy is used in wind-up toys, in springs and, as we have seen, bungee cords. It is also elastic energy that enables you to bounce a basketball or a tennis ball, and is the energy stored in an archer's stretched bowstring (Figure 8.2.5).

▲ **FIGURE 8.2.5**   Potential energy is stored energy that has the potential for later use. The archer pulls back on the bowstring and the elastic energy it holds is released to give the arrow the energy to move.

Mechanical energy

Mechanical energy is the energy that is possessed by an object due to its motion or due to its position. Mechanical energy can be either kinetic energy (energy of motion) or potential energy (stored energy). Some objects possess kinetic and potential energy, with their mechanical energy being the sum of both. For example, when you are at the top of a loop on a roller coaster, you have kinetic energy due to the movement of the carriage and gravitational potential energy due to your altitude.

mechanical energy
the sum of an object's kinetic and potential energy

8.2 LEARNING CHECK

1 **List** three objects that possess potential energy.

2 **Describe** two types of energy that are classified as potential energy.

3 Gravitational potential energy plays an important role in many professional sporting events. Look at Figures 8.2.6 and 8.2.7. **Explain** why the landing area of a pole vault has a lot more cushioning than for a high jump.

▲ FIGURE 8.2.6   High jump

▲ FIGURE 8.2.7   Pole vault

4 **Conduct** a plus, minus, neutral analysis of the benefits of food labels on different types of packaged foods.

5 **Explain** why a netball that has been thrown into the air possesses mechanical energy.

Video activity
Rube Goldberg machine

Interactive resources
Simulation: Energy transfers
Label: Energy transfer diagrams

GET THINKING

Energy is more useful in our lives when we can transfer energy between objects; for example, the transfer of heat that takes place when cooking. As you complete this module, think about other examples of energy transfers that take place within your home or at school.

Everyday energy transfers

When cooking, a source of energy such as natural gas is used to produce heat, which is transferred to the cooking utensil and the food that is being cooked. But what do we mean by **energy transfer**? Energy occurring in one place can be passed on as the same type of energy to another place. If you kick a football and it moves, some of your foot's kinetic energy is transferred to the ball to make it move. When a kettle of water is boiled, heat is transferred from the heating element to the water. These are examples of everyday energy transfers (Figure 8.3.1).

energy transfer
the movement of a single type of energy from one place to another or from one body to another

▲ **FIGURE 8.3.1** Energy can be transferred from one place to another.

You should now also be familiar with types of kinetic and potential energy. In this module, we will explore how these types of energy are transferred.

Law of conservation of energy

law of conservation of energy
a law of physics that states that energy cannot be created or destroyed

Before going into energy transfer in more detail, let's briefly look at the **law of conservation of energy**. This law states that, in any isolated system, energy cannot be created or destroyed. This indicates that the total amount of energy within any isolated energy system must remain constant. You will investigate this important law of physics in more detail in Year 9.

Examples of transferring energy

Energy, in its various forms, may be transferred from one object or location to another. The scientific way of describing energy being transferred is *doing work*. Heat is often the form that energy takes when being transferred. Table 8.3.1 provides some examples of what is needed to transfer energy and some simple examples.

▼ **TABLE 8.3.1** Examples of energy transfer

Energy transfer	Examples
Mechanical, involving a force	• On a pool table, a moving white ball collides with and transfers some of its kinetic energy to a stationary red ball, forcing it to move.
Electrically, involving an electric current	• An electric current from a power point is transferred through a charger to the battery of a computer or tablet.
Using light or sound waves	• Rays of light from the Sun are absorbed by a solar hot water panel to heat water. • Sound energy is transferred from a guitar to your ear as you hear it being played.
Heating	• A kettle of water is boiled by a heating element. • A reverse-cycle air conditioner is used to heat a living room. • A heat lamp is used to keep food hot in a servery.

Transferring kinetic energy

When objects collide, there is a transfer of energy from one object to another. In the example of a footballer kicking a football, some of the kinetic energy in their foot is transferred to the football as their foot contacts the ball. This is an example of a mechanical transfer of energy. Many sports rely on the transfer of kinetic energy from one object to another. When an object such as a bat strikes a ball, there is a transfer of kinetic energy from one object to another: from the bat to the ball (Figure 8.3.2).

▲ **FIGURE 8.3.2** A cricketer transferring kinetic energy from bat to ball

▲ **FIGURE 8.3.3** Newton's cradle – an example of transferring kinetic energy

Can you think of some other sport examples where a force transfers kinetic energy? Other common examples of transferring kinetic energy include accidents, such as when cars collide or when you accidentally knock over an object.

In a Newton's cradle, shown in Figure 8.3.3, the kinetic energy of the end sphere is transferred to the sphere it hits, then continues to be transferred between the stationary spheres until it reaches the sphere at the opposite end, causing it to move. This is another example of a mechanical transfer.

Sound is another example of transferring energy as kinetic energy. We are able to hear music or people talking because sound energy has transferred from the source of the sound to our ears (Figure 8.3.4).

▲ **FIGURE 8.3.4** Sound energy is transferred when we (a) hear music being played or (b) hear people talking.

▲ **FIGURE 8.3.5** Transferring electrical energy to recharge an electric vehicle

Transferring electrical energy

Electrical energy is another type of energy that is often involved in energy transfers. The transfer takes place by moving charges in electric circuits. All your electrical devices start as a transfer of electrical energy from a power source – which may be a power point or battery – through wires. Distributing electricity from a power station to where it is used – for example, to our houses – requires electrical energy to be transferred. Electric vehicles rely on the transfer of electrical energy to recharge their batteries (Figure 8.3.5).

9780170463027

Modelling energy transfer with diagrams

Scientists often use diagrams to represent the flow of something. When discussing energy, the flow of energy can be represented using a flow chart. In a flow chart, the arrow shows the direction of the energy transfer from the source of the energy to the receiver. We will use the example of cooking food (Figure 8.3.6). An arrow shows the flow of the input energy, a box contains the receiver of the energy, and an arrow is pointing to the energy output.

Consider another example of wind moving a yacht. The kinetic energy of the wind is captured by the sail to transfer kinetic energy to the yacht (Figure 8.3.7).

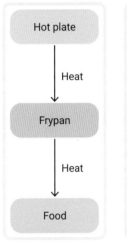

▲ FIGURE 8.3.6
Example of a simple energy transfer diagram

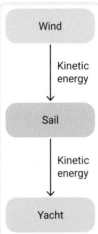

▲ FIGURE 8.3.7
The transfer of kinetic energy from the wind to the sail of a yacht

8.3 LEARNING CHECK

1 **Describe** energy transfer by using an example.
2 **Identify** the energy transferred from a lit gas hot plate to a saucepan of water.
3 **State** the law of conservation of energy.
4 **List** three common devices in your household that require a transfer of electrical energy.
5 Construct an energy transfer diagram for a hockey player hitting a ball.
6 **List** three common devices in your household that transfer energy mechanically using a force.
7 **Explain** why energy is more useful to us when it is transferred between objects.
8 **Describe** how a transfer of kinetic energy can be:
   a helpful.
   b harmful.

BY THE END OF THIS MODULE, YOU WILL BE ABLE TO:

✓ describe how energy can be transformed through energy systems

✓ use flow diagrams to model energy transformation.

Video activity
Energy transformation

Interactive resource
Drag and drop: Renewable or non-renewable?

Extra science investigations
Transferring and transforming energy
Transforming gravitational energy

energy transformation
to change one type of energy into another type of energy

GET THINKING

We rely on energy transformations to make use of different sources of energy to do work in quite different situations. For example, electric vehicles would not move unless the chemical potential energy stored in their batteries was transformed into electrical energy, which is then transformed into kinetic energy. As you complete this module, imagine what it would be like if energy transformations were not possible in your daily life.

Transforming energy

Transforming energy means to convert one type of energy into another type of energy. We rely on **energy transformations** in all aspects of our life. One important example of energy transformation you learned about in Chapter 3 is when green plants transform light energy into chemical potential energy by photosynthesis. Our bodies use the chemical energy stored in the food we eat to keep us alive, by transforming it into other types of energy.

Many devices and appliances transfer and transform energy. In many instances, there will be both energy transfer and energy transformations occurring, especially in complex energy systems. Table 8.4.1 provides some examples of energy transformations.

▼ TABLE 8.4.1 Simple examples of energy transformations

Example	Energy transformation(s) involved
Burning a log	• Chemical energy stored in the wood transforms into heat and light energy.
Turning on a torch	• Chemical energy stored in the battery transforms into electrical energy and then into light energy in the bulb. • When left on, the bulb will get hot, showing that some electrical energy has transformed into heat energy. • When an electric current flows, some heat is generated by the friction of the charges moving through the wires.
Knocking a glass off a table	• Gravitational potential energy is transformed into kinetic energy as the glass falls. • Some kinetic energy is transformed into sound energy when the glass hits the floor and shatters.

Modelling energy transformation with diagrams

In Module 8.3, you were introduced to flow charts to represent energy transfers. We can use similar diagrams to represent energy transformations. Arrows show the direction of energy flow within the system, and the energy types are shown in the order in which transformations take place.

Let's use the example of turning on a torch to construct a flow diagram (Figure 8.4.1).

When considering energy transformation diagrams, we are interested in determining the order in which transformations of energy take place. The flow chart is simplified into a chain of connecting energy types (Figure 8.4.2).

Transforming energy into electrical energy

Electrical energy is one of the most common types of energy we depend on every day. We use electricity in our homes, at work or at school, in industry and in many other places. Today's society relies heavily on the electricity generated using various energy sources.

Non-renewable energy sources

Non-renewable energy sources include fossil fuels such as coal, oil and natural gas. Some countries also use nuclear power. Coal has been an important energy source for electricity generation in Australia for many years. Using fossil fuels to produce electrical energy in power stations relies on transforming their chemical potential energy into heat by burning the fuel. The heat is then used to convert water into high-pressure steam. The steam drives a turbine by kinetic energy to transform mechanical energy (the combination of kinetic and potential energy) into electrical energy (Figure 8.4.3).

There is now a significant push from the community to shift away from coal and other fossil fuels, and to instead use energy sources that produce fewer **greenhouse gases** and reduce the impact on global warming.

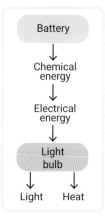

▲ **FIGURE 8.4.1**
Energy transformations in a torch

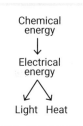

▲ **FIGURE 8.4.2**
A flow chart for chemical energy being transformed into light and heat

non-renewable energy source
a source of energy that is finite in nature; i.e. used at a faster rate than it can be produced

greenhouse gases
heat-trapping gases, such as carbon dioxide, that have been linked to global warming

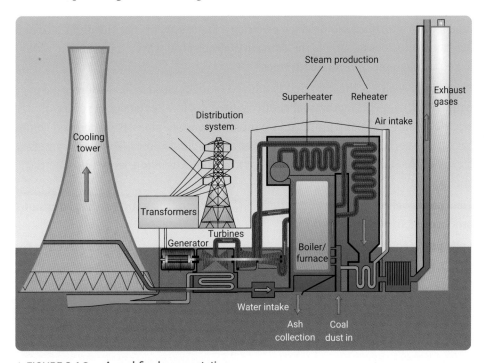

▲ **FIGURE 8.4.3**   A coal-fired power station

Renewable energy sources

Renewable energy sources are sources of energy that can be produced at a faster rate than they are used; they can be 'renewed'.

- **Biofuels**, like fossil fuels, are burned to release heat. The advantage of biofuels is that the fuel is made from recycled biological material that would normally be waste material. The energy captured in these materials is used as a fuel rather than ending up being incinerated or breaking down in the environment.

- In the production of **hydroelectricity**, gravitational potential energy is used. Water drops from a height, transforming the gravitational energy into kinetic energy that drives the turbine (Figures 8.4.4 and 8.4.5).

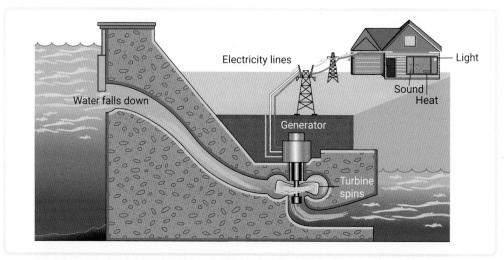

▲ **FIGURE 8.4.4**   Transforming gravitational potential energy into electrical energy using a hydroelectric power station

- **Tidal energy** can be used to generate electricity. The kinetic energy of moving water is used to turn a turbine and produce electricity. Some systems use the difference in the heights of the tides to act like the dams in hydroelectric power stations. The water is forced through pipes to rotate turbines.

- **Geothermal energy** is the heat that is trapped in rocks close to heat sources deep within Earth's crust. Geothermal energy is used in New Zealand as a source of electricity. Water, usually pumped into these regions via pipes, is heated and comes back to the surface as steam, under pressure, in separate pipes. The significant kinetic energy of the steam drives a turbine to transform the kinetic energy into electrical energy.

- **Wind power** generators transform the kinetic energy of the wind to drive a turbine to produce electricity.

- **Solar power** relies on converting light energy from the Sun directly into electrical energy in solar cells.

Sidebar definitions

renewable energy source
a source of energy that can be produced at a faster rate than it can be used

biofuels
fuels that are made from biological sources that can replace fossil fuels

hydroelectricity
electrical energy produced by transforming gravitational energy of falling water into kinetic energy to drive a turbine

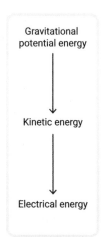

▲ **FIGURE 8.4.5**
A simple energy transformation flow chart for generating hydroelectricity

tidal energy
electricity generated from the ebb and flow of the tides

geothermal energy
heat that is trapped in rocks close to heat sources deep within Earth's crust

wind power
electricity generated by harnessing the kinetic energy of the wind to drive a turbine

solar power
electricity generated directly from sunlight, using solar cells

9780170463027

Waste energy

In any transformation of different energy types into electrical energy, there will always be **friction** due to the movement of electric charges in the wires and the moving parts of the electrical generating system. This results in some of the available energy being transformed into heat. This heat is **waste energy** because it is not usable for the main purpose of the system, which is to generate electricity.

Transforming energy by friction

When contacting surfaces rub up against each other, frictional forces are always present. As we have learned, this friction transforms some of the kinetic energy of the surfaces rubbing together into heat. When this heat is waste energy, it may be a problem. For example, the pistons in a car engine require some method of removing the heat or reducing the friction; otherwise, the engine will be damaged (Figure 8.4.6). One way to reduce friction is to use a lubricant, such as oil or grease.

Heat produced through friction can also be helpful. Rubbing your hands together transforms stored chemical energy in your muscles into kinetic energy and the friction between the surfaces of your hands produces heat. The temperature of your hands will increase. Many cultures have used friction to produce heat to start fires, including First Nations Australians and other indigenous populations around the world.

8.4

friction
the force that resists movement when one surface slides over or rubs against another

waste energy
energy that is a result of energy transformation, but that is not usable within the energy transformation system

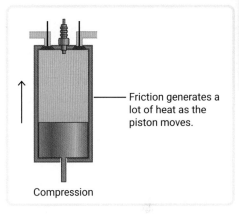

Friction generates a lot of heat as the piston moves.

Compression

▲ **FIGURE 8.4.6** Friction generates heat in a moving engine piston.

☆ ACTIVITY

Creating energy flow charts

What to do

Construct energy flow charts for the energy transformations taking place in each of the following situations.

1 Paddling a canoe
2 Using a computer or mobile phone to play a computer game
3 Swimming
4 A plane taking off or landing
5 Using a water slide at a theme park
6 Surfing

8.4 LEARNING CHECK

1 Using a suitable example, **define** 'energy transformation'.
2 **Compare** the types of energy involved in a bungee jump from when the person first jumps until they are pulled back up.
3 **Explain** the difference between energy transfer and energy transformation, using a suitable example.
4 **Construct** a simple energy flow chart to show the energy transformations involved in pedalling a bicycle up a small hill and then going down the other side without pedalling.
5 Using examples, **explain** why heat is produced in many different energy transformations.

8.5 First Nations Australians' traditional fire-making techniques

IN THIS MODULE, YOU WILL:

✓ examine how First Nations Australians transform movement energy into heat to make fire.

Traditional tools and methods of fire starting

Prior to the invention of the match in the early 1800s, there were several different methods used for making fire. While in many cultures only one method was used, across Australia, First Nations Australians knew and used four different methods. These are: the hand drill, the fire saw, the fire plough and the percussion methods.

In the first three methods, heat is generated by rubbing two pieces of wood together to produce a glowing ember, which is then used to start a fire. In the percussion method, a piece of flint, resting on tinder (dry material such as grass, that easily catches fire), is struck in glancing blows with a piece of iron pyrite. The striking causes a spark that then lands on and ignites the dry tinder.

The hand drill and fire saw were the most commonly used methods across Australia. The fire plough method was mainly used in north-western Australia, while records indicate use of the percussion method by First Nations Australians groups in South Australia and Tasmania.

▲ FIGURE 8.5.1   In the hand drill method, friction produces enough heat to produce a glowing ember. This drill stick and hearth is from south east Queensland and is made from grass tree and hibiscus.

The hand drill method

The hand drill method (also referred to as the fire drill or drill stick method) was most commonly used across coastal and northern areas of Australia.

First Nations Australians require the following materials for the hand drill method:

- a drill stick – a long (around 70 cm), thin, straight piece of stick, with one rounded end
- a hearth – a flat piece of wood with an indentation that has a side notch
- tinder – easily combustible material such as small pieces of dried grass or leaves, shaved wood or dried kangaroo dung.

The hearth is placed on the ground and tinder is put under the side notch. The rounded end of drill stick is placed in the indentation in the hearth and twirled vigorously between the hands while applying a downward pressure. The friction between the two pieces of wood produces fine sawdust and heat. The heat causes the sawdust to smoulder, and this ignites the tinder.

The type of timber used for the drill stick and the hearth depended on what was available in First Nations Australians' Country/Place. Softwood of the same type was

9780170463027

often preferred for both parts. Less energy was required to produce an ember because of the low density and low heat conductivity of softwood. When two different types of wood were used, the drill stick was usually made of the harder wood.

1 **Explain** where heat comes from in the hand drill method.

2 Draw an energy flow chart to show the energy transformations that happen using the hand drill method.

3 **Explain** why less energy is needed to produce an ember when using softwood than when using hardwood.

The fire saw method

The fire saw method was mainly used by First Nations Australian groups throughout inland Australia and the north-west coastal area.

First Nations Australians require the following materials for the fire saw method:

- a fire saw – a piece of hardwood with a sharp edge; often, available objects such as a boomerang, spear thrower, coolamon or specially produced wooden knife were used as the fire saw
- a hearth – a softwood shield with a groove, or a split piece of wood held open by wedges (sometimes called a cleft stick)
- tinder – easily combustible material such as small pieces of dried grass or leaves, shaved wood or dried kangaroo dung.

▲ FIGURE 8.5.2 (a) A cleft stick fire saw hearth (unknown origin) and (b) a Central Desert spear thrower being rubbed in a sawing motion across a shield as the hearth

The hearth is placed on the ground. Pieces of tinder are placed in or around the split or groove. The fire saw is placed across the hearth, resting in a notch. The saw is vigorously pulled back and forth using a sawing motion. As with the hand drill method, sawdust and heat are produced, causing the tinder to catch alight.

☆ ACTIVITY 2

1 What type of timber would be the best to use for the hearth? Why?

2 **Draw** an energy flow chart for the fire saw method.

3 a **Compare** the hand drill and fire saw methods.

   b Identify which method you think would be the most energy efficient and **explain** why.

4 Figure 8.5.3 shows a demonstration of the fire plough method. Figure 8.5.4 shows a close-up of the fire plough. Using what you have learned, **suggest** materials and a technique for this method.

▲ FIGURE 8.5.3   The fire plough method

▲ FIGURE 8.5.4   A close-up view of a fire plough

5 The percussion method involves using a piece of flint and iron pyrite to generate a spark. **Compare** this method to the hand drill and fire saw methods in terms of energy changes.

9780170463027

8.6 Generating electricity in remote communities

BY THE END OF THIS MODULE, YOU WILL BE ABLE TO:
✓ explain how remote communities generate electricity.

Vast areas of inland Australia do not have access to the same electricity sources that are available in metropolitan areas. Communities in these remote areas rely on small electrical generation systems to supply their local electricity needs. Many of these communities, and particularly isolated properties, have relied or still rely on diesel generators as their source of energy. These generators are expensive to run continuously. They require a lot of maintenance and use a lot of diesel fuel.

Video activity
Clean energy in remote communities

Many isolated communities are looking to use remote area power supplies (RAPS) to provide low-cost sustainable energy to meet their needs – stand-alone energy-generating systems that are not connected to a main electricity grid. The type of energy-generating system that can be used depends on where the communities are located, the size of the communities and the available energy resources to drive the system. Because of the isolation of remote communities, RAPS are suited to using renewable energy resources such as solar, wind or moving water, and the more common generating systems in use include solar-powered photovoltaic cells, wind turbines and hydropower generators (Figure 8.6.1).

▲ **FIGURE 8.6.1** A hybrid RAPS. It uses energy from wind and moving water to turn generators that produce electric power, as well as solar-powered photovoltaic cells. The back-up is a motor generator powered by fossil fuel.

Many remote communities use several interconnected hybrid systems. The energy generated by these systems is used directly and excess energy is stored in special batteries. These systems are usually backed up by generators that run on fossil fuels.

8.6 LEARNING CHECK

1 **List** the factors to consider when determining the likely energy sources that an isolated or remote community could use to generate electricity.
2 **Analyse** the benefits and challenges of using an integrated hybrid system in a remote or isolated community.
3 If you were the owner or manager of a small, remote ecotourism resort (five staff members and a maximum of 25 guests), **list** the things that you would put in place to use the available electrical energy efficiently and wisely. Briefly account for each of your ideas.

Modelling energy transfers and transformations with flow charts

SCIENCE SKILLS IN FOCUS

IN THIS MODULE, YOU WILL FOCUS ON LEARNING AND IMPROVING THESE SKILLS:

▶ use and construct diagrams to model data

▶ use diagrams to communicate information.

In many disciplines of science, diagrams and flow charts are used to provide instruction on how to complete something, such as assembling an apparatus to perform an investigation. Diagrams are also used to describe observations and to model data. In this investigation, you will use energy flow charts to describe your observations and to model the flow of energy in a series of simple activities.

Video
Science skills in a minute: Organising data into charts

Science skills resource
Science skills in practice: Energy transfers and transformations

▶ **To construct an energy flow chart, you need to:**

• identify the initial source of energy and the relevant type of energy the source provides

• make careful observations to identify any energy transfers and transformations taking place

• construct a flow diagram labelling the initial type of energy and using arrows to show the direction of any energy transfers or transformations in the order in which they occur.

In the case of an energy transfer, a box should be drawn around the object receiving and then transferring the energy.

If several energy transformations occur at the same time from one type of energy, the arrows should indicate this.

MODELLING ENERGY TRANSFERS AND TRANSFORMATIONS

AIM

To model energy transformations and transfers using flow charts

YOU NEED

☑ 18 alligator clips
☑ tennis ball
☑ buzzer
☑ 3×1.5-volt DC batteries
☑ 9 connecting wires
☑ filament light bulb in light bulb holder
☑ resistance wire
☑ 3 switches
☑ wind-up toy car with an on/off switch

 Warning

Be careful when handling any equipment that may have sharp edges. Be mindful of any objects that may become hot in this activity. Some may be hot enough to burn your fingers if you touch them for too long.

WHAT TO DO

1 Construct a suitable table with three columns with headings: 'Object', 'Initial energy type' and 'Observations'.

2 Your teacher may have set up five stations to complete this investigation. If so, move around from Station 1 to Station 5.

Station 1 Drop the tennis ball from shoulder height. Record your observations.

Station 2 Connect the battery to a switch and the buzzer using the connecting wires and alligator clips (Figure 8.7.1). Press the switch. Record your observations.

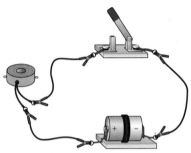

▲ FIGURE 8.7.1

Station 3 Wind up the toy car, making sure the switch is in the 'off' position. Record your observations. Turn the switch of the wind-up toy car to 'on' and place it on the bench or floor. Record your observations.

Station 4 Connect the battery to a switch and the resistance wire (Figure 8.7.2). Feel the temperature of the wire. Press the switch for several seconds and quickly feel the temperature of wire (be careful). Record your observations.

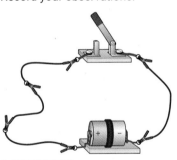

▲ FIGURE 8.7.2

Station 5 Connect the battery to the light bulb and switch (Figure 8.7.3). Feel the temperature of the light bulb. Press the switch for several seconds and feel the temperature of light bulb (be careful). Record your observations.

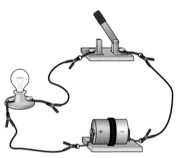

▲ FIGURE 8.7.3

WHAT DO YOU THINK?

1 Identify which (if any) of the five activity stations are examples of:
   a energy transfer.
   b energy transformation.

2 What criteria did you use to distinguish energy transfers from energy transformations in question 1?

3 Identify any of the situations that have a combination of energy transfer and energy transformation.

CONCLUSION

Construct energy flow charts for each of the five stations. Remember that some of the actions may involve more transfers and/or transformations than others.

8 REVIEW

REMEMBERING

1 **Define** 'kinetic energy'.

2 **Define** 'potential energy'.

3 How do scientists define energy?

4 **Name** the metric and imperial units we use to measure energy.

5 **Identify** the following sentences as true or false. If they are false, rewrite them to make them true.

a Energy is described by where it comes from or how we experience it.

b Cold objects transfer heat to hot objects.

c Fossil fuels are sources of thermal energy.

d An avalanche transforms gravitational potential energy mostly into sound energy.

e Boiling water using an electric kettle is an example of an energy transfer.

f Transferring energy and transforming energy mean the same thing.

UNDERSTANDING

6 Why do we classify chemical energy as a type of potential energy, but we classify heat as a type of kinetic energy?

7 **Name** the type of energy stored in a battery.

8 **Construct** a flow chart to show the energy transformations in producing electricity using fossil fuels.

9 **Name** three objects that store elastic potential energy.

APPLYING

10 **Identify** three sources of potential energy stored in your house as:

a chemical potential energy.

b elastic potential energy.

c gravitational potential energy.

11 **Explain** why the sources you identified in Question 10 are examples of stored energy.

12 **Explain** why charging an electric vehicle is an example of an energy transformation. Justify your explanation with a flow chart.

13 Fuels are chemicals that can be used to produce electrical energy. **Identify** at least four fuels that are used for this purpose. Explain the energy transformations required to use each of these fuels to generate electricity.

14 **Name** two different types of energy involved in:

a playing a sport.

b recreational activities other than sport.

15 **Construct** a flow chart that demonstrates a type of energy you identified in Question 14a being transformed into other types of energy.

16 **Construct** a flow chart that demonstrates a different type of energy you identified in Question 14b being transferred between objects.

17 In the action of throwing a netball, a person transfers energy but does not transform energy. Is this statement true or false? **Explain** your answer.

18 **Explain** the energy changes for the fire plough method of fire-making used by First Nations Australians.

EVALUATING

19 **Explain** why heat is produced when electrical energy is transformed into light energy in a light globe.

20 A bungee jumper has maximum gravitational potential energy just before they jump. As they fall, gravitational energy transforms into kinetic energy. The jumper slows down as the elastic rope begins to stretch. When they reach the bottom of their jump, the rope pulls them back up.

a **Explain** why the bungee jumper slows down as the elastic rope is stretched.

b **Explain** why the bungee jumper has maximum gravitational potential energy before jumping.

21 **Construct** an energy flow chart to explain the energy transformations of the bungee jumper in Question 19.

22 **Describe** how we can reduce the amount of 'wasted energy' in a house being heated during winter.

23 **Explain** why sound energy requires a material for the energy to be transferred.

CREATING

24 **Construct** a simple mind or concept map of the key words from the chapter. Make as many links as possible between them.

BIG SCIENCE CHALLENGE PROJECT #8

1 Connect what you have learned

In this chapter you have learned about types of energy and how they are classified as types of kinetic or potential energy. You have seen examples of energy transfers, where the same type of energy is transferred from one object to another. You have also learned about transforming energy from one type to another. The Rube Goldberg machine shown below contains examples of energy transfers and transformations. Many of the transfers involve transferring kinetic energy from one object to another.

2 Check your thinking

Look carefully at the Rube Goldberg machine. This example involves the use of common items such as balls, rope, pulleys, pavers and wooden ramps, as well as some less common and strange items, to make a complex machine. The machine uses a complicated mix of energy transfers and energy transformations to accomplish a task. It is a very complicated way of completing a simple task, but can be a lot of fun!

3 Make an action plan

Plan your own Rube Goldberg machine individually or as part of a team. The machine must contain at least five steps to complete the simple task of pouring water into a pet's bowl. The machine must include at least two energy transfers and at least three energy transformations. In your planning, you (or your group) may be as creative as you like, but your machine should use common household items. The plan must include an energy flow chart to show the energy transfers and transformations that take place to complete the task.

4 Communicate

Draw a plan for your Rube Goldberg machine. Write a set of instructions, including energy flow charts, to explain how the machine works to complete the task. Construct the machine and film it in operation, explaining the energy transfers and transformations as they occur. Alternatively, you may wish to construct the machine and demonstrate it to other students in your class.

Note: Constructing and filming the Rube Goldberg machine is not part of the assessment.

GettyImages/Stone/Jeffrey Coolidge

Glossary

A

absolute dating using scientific equipment to determine how old something is

adult stem cell an undifferentiated cell found everywhere in the body, which can divide to replace dead or damaged cells

alveoli (singular: alveolus) air sacs at the end of the bronchioles in the lungs

amylase the enzyme that digests carbohydrates

anus the external opening of the rectum, through which faeces leave the body

aorta a large artery in humans that takes blood from the left side of the heart to the body

appendix a small tube-shaped sac attached to, and opening into, the lower end of the large intestine

artefact an object made by a human being

artery a blood vessel in humans that carries blood away from the heart

asthenosphere the solid part of the mantle below the lithosphere that can flow

atherosclerosis the build-up of deposits on the inner wall of arteries

atmosphere the gaseous layer surrounding Earth

atom the smallest part of an element that contains the properties of that element

atomic weight the mass of an atom on Earth

atria (singular: atrium) the two upper chambers in the human heart, which receive blood from veins

axon the part of a neuron that carries the nerve impulse

B

batholith a very large volume of intrusive rock, formed deep under Earth's surface by solidification of magma

bed a horizontal layer of sedimentary rock

biconcave shaped like a flattened disc with dips on both sides

biodegradable describes a substance or product that is able to completely break down and return to natural products within a short time

biodiversity the variety of living species on Earth, including plants, animals, bacteria and fungi

biofuels fuels that are made from biological sources that can replace fossil fuels

biological weathering a process of weathering in which living organisms break down rocks

biosphere the parts of Earth where life is found

body fossil a fossil formed from the remains of a plant or an animal

boiling changing state from a liquid to a gas at a rapid rate

bolus a ball of food that passes into the oesophagus from the mouth

bowel motion the process of egesting faeces through the anus

bronchi (singular: bronchus) tubes that branch off the trachea to the left and right lung

bronchiole a smaller tube made when the bronchi divide

C

caecum a pouch or large tube-like structure at the beginning of the large intestine; receives undigested food material from the small intestine

calorie (cal) the imperial unit of measurement for energy

canopy the 'top' of a forest, made up of overlapping leaves and branches of tall trees

capillary a very small blood vessel in humans, located in between the smallest arteries and smallest veins

carbohydrates a complex food group found in starchy foods such as bread and rice

carbon sink a natural or artificial storage place for carbon, for an indefinite period

carnivorous describes an organism that feeds on animals only

cartilage flexible tissue that makes up part of the skeleton

cast a fossil formed when minerals fill a fossil mould

9780170463027

cell the basic structural unit of all living things

cell theory the basic theory in modern cell biology that states that all living things are made up of cells, cells are the basic units of all living things, and cells form from existing cells

cell wall the rigid outer covering of bacterial and plant cells that surrounds the cell membrane

cellular respiration a series of chemical reactions that break down glucose into chemical energy

cellulose a complex carbohydrate that makes up the cell walls of plants

cementation the process whereby new minerals bind sediment grains together

chamber one of the compartments that form the structure of the heart; there are four chambers in the human heart

chemical bond a force that holds atoms together

chemical change when the chemical make-up of a substance changes, and a new substance or substances are formed

chemical digestion the chemical breakdown of food into simpler substances

chemical equation a symbol summary that shows the reactants and products of a chemical reaction

chemical formula a collection of symbols and numbers that represent the number of atoms in a molecule or compound

chemical potential energy energy stored in the chemical bonds of a substance and released when the substance reacts

chemical properties the properties of a substance that determine how it reacts when combined with other substances

chemical reaction a process that occurs when a substance changes to produce a new substance

chemical symbol a letter or letters of the Latin alphabet used to represent an atom of a specific element

chemical weathering a process of weathering that changes the chemical composition of the minerals in rocks

chlorophyll the green pigment in chloroplasts that absorbs light for photosynthesis

chloroplast a green organelle in plant cells that contains chlorophyll and carries out photosynthesis

cholesterol an insoluble, waxy substance

chyme partially digested food that passes from the stomach to the small intestine

clastic small pieces of rock or mineral

cleavage the way a mineral splits to produce a flat surface

coarse focus knob a knob that adjusts a microscope so that it focuses on the specimen by rapidly raising and lowering the stage

colloid a mixture of two or more insoluble substances that remains evenly mixed and does not settle over time

compaction the process whereby pressure forces particles closer together

companion cell in plants, a cell adjacent to a sieve tube cell, which makes substances unable to be made by the sieve tube cell

compound a chemical substance made up of two or more different atoms chemically bonded together; bonded atoms can be a metal and a non-metal (ionic compound), or two non-metals (molecular compound)

concentration the amount of solute present in a specified amount of solution

condensing changing state from a gas to a liquid

contact metamorphism rock metamorphism caused by heat

continental drift the movement of the continents across the surface of Earth over geological time

continental plate a lithospheric tectonic plate containing a lot of continental crust

convection the process that transfers heat in a fluid

convergent boundary the border at which crust is destroyed as one plate moves beneath another, or where two continental plates collide

corrode the breakdown of a metallic substance due to a chemical reaction with chemicals in the environment, such as oxygen or water

crust the outermost and thinnest layer of Earth

crystal a solid in which the atoms are arranged in a well-ordered pattern

cuticle the waxy protective layer on the surface of a plant

cytoplasm a jelly-like substance that fills the inside of a cell

daughter cells two cells that result from the division of a parent cell

density the mass of a substance in a specific volume

deposition the laying down of sediment

diagnosis the identification of the nature of an illness

diaphragm a sheet of muscle under the lungs that assists with inhalation and exhalation

diatomic a molecule consisting of two atoms (*di*– means 'two'; *atomic* means 'atom')

differentiation a biological process whereby cells of an organism become specialised

diffusion the movement of gas or liquid particles from a region of high concentration to a region of low concentration

divergent boundary the border at which new crust is formed as tectonic plates pull away from each other

earthquake a violent shaking of the ground caused by energy-carrying waves

egest to pass out of the body

elastic potential energy energy stored in a spring or elastic material

electrical energy energy carried by charged particles in electric circuits

electron a type of negatively charged particle

element a pure substance made up of only one type of atom; cannot be broken down into a simpler substance

embryonic stem cell a cell from an embryo that is three to five days old; can divide into more stem cells or can become any type of cell in the body; it is pluripotent

endoplasmic reticulum an organelle that consists of an interconnecting system of thin membrane sheets dividing the cytoplasm into compartments and channels

endoscope an instrument used to look inside the human body

endothermic reaction a chemical reaction that takes in heat energy

energy the ability to do work, such as moving or cooking, or the potential to do work, such as the chemical potential energy stored in food

energy transfer the movement of a single type of energy from one place to another or from one body to another

energy transformation to change one type of energy into another type of energy

enzyme a protein found in the body that speeds up a chemical reaction

ephemeral describes a plant completing its life cycle quickly and spreading large quantities of seeds

epicentre the point on Earth's surface directly above an earthquake focus

epidermis the cellular surface layer of a plant

erosion the movement of weathered material away from where it forms by water, wind, ice or gravity

estuary an area where a freshwater river meets the ocean

eukaryote an organism composed of one or more cells that contain a nucleus and membrane-bound structures

evaporating changing state from a liquid to a gas

excretion the process of eliminating or expelling waste matter

exothermic reaction a chemical reaction that releases heat energy

expiration breathing out

extrusive describes an igneous rock formed from lava at or above the surface of Earth

eyepiece lens a lens on a microscope through which the eye views the image formed by the objective lens

faeces undigested waste material

fats a complex food group found in foods such as butter and cream

fault a fracture on Earth's surface where rocks have moved due to tension or compression forces

fibre the indigestible parts of plants

fibrous roots many small roots of similar size that grow from the bottom of the stem of some plants

field of view the diameter of the circular area that appears when you look into a microscope

filtration the process in the kidney where all materials, except for protein and blood cells, are forced out of the bloodstream

fine focus knob a knob that adjusts a microscope so that it focuses on the specimen by slowly raising and lowering the stage

fissure a long fracture or crack

flammability the ability of a substance to catch fire

focus the place where earthquake energy is released

foliation layering in a rock formed by crystal regrowth

fossil the remains or traces of living organisms, preserved in rock

freezing changing state from a liquid to a solid

friction the force that resists movement when one surface slides over or rubs against another

gangue minerals minerals found in rocks that are less valuable than ore

geosphere the solid part of Earth

geothermal energy heat that is trapped in rocks close to heat sources deep within Earth's crust

Golgi body an organelle that processes, packages and stores proteins and lipids

gravitational potential energy energy associated with the position of an object above a reference height

greenhouse gases heat-trapping gases, such as carbon dioxide, that have been linked to global warming

group (in chemistry) a vertical column in the periodic table

guard cells cells that surround the stomata of a plant, allowing them to open and close

haemoglobin the component of red blood cells that binds with oxygen

hard parts refers to the hard parts of an organism's body, such as shells, bones and teeth

heartburn a burning feeling in the oesophagus caused by rising stomach acid

heat thermal energy of an object that can be transferred to another object

herbivorous describes an organism that feeds on plants only

hormone a chemical messenger

hotspot an unusually hot area in Earth's upper mantle where the mantle melts

hydrochloric acid a type of acid; in the stomach it helps to digest food

hydroelectricity electrical energy produced by transforming gravitational energy of falling water into kinetic energy to drive a turbine

hydrosphere all the water on Earth

hydrothermal metamorphism rock metamorphism caused by hot fluids in the earth changing the chemical composition of rocks they pass through

igneous rock a type of rock formed when molten materials (magma or lava) cool and solidify

immune system a complex system that defends the human body against infection and disease

induced pluripotent stem cell a cell that is reprogrammed from adult tissues to become pluripotent

ingested taken in; eaten

inner core the innermost solid, metal part of Earth

inorganic a substance not formed from the remains or products of living things

insoluble not able to be dissolved

inspiration breathing in

intrusive describes an igneous rock formed from magma below the surface of Earth

ionic compound a compound formed when a metal and a non-metal chemically bond

iris diaphragm a part of a microscope that regulates the amount of light that strikes the specimen

joule (J) the metric unit of measurement for energy

kidneys the excretory organs of mammals

kinetic energy the energy possessed by an object or particle that is in motion

lava hot, molten rock that is expelled during a volcanic eruption

law of conservation of energy a law of physics that states that energy cannot be created or destroyed

light energy energy transferred as a form of electromagnetic radiation

light microscope a microscope that uses light to view the specimen

lignin a material that stiffens and strengthens plant cell walls

lignotuber a partly underground swelling of the trunk of a plant, with many buds that sprout after fire

lithification the process whereby a sediment under compaction becomes a rock

lithosphere Earth's crust and the rock-like upper part of the mantle

lustre how light looks when reflected from the surface of a crystal, rock or mineral

lysosome an organelle that breaks down and recycles old, worn-out cell organelles

magma extremely hot liquid or semi-liquid rock formed under the surface of Earth

magnetic reversal a change in the direction of Earth's magnetic field

magnification the action of enlarging the apparent size of a specimen being observed

mantle the rock layer between the crust and the core of Earth

mechanical digestion the physical breakdown of food into smaller pieces

mechanical energy the sum of an object's kinetic and potential energy

melting changing state from a solid to a liquid

membrane the thin layer that forms the outer boundary of a living cell, or of an internal cell compartment

meniscus the downward or upward curve at the top of a liquid caused by surface tension and adhesion between the liquid molecules and the container

metal a chemical element that has certain properties, such as conducting heat and electricity, being malleable and being ductile

metalloid a chemical element that has properties in between those of a metal and a non-metal

metamorphic describes rock that has been altered by heat, pressure or hot fluids

micrograph a photograph taken using a microscope

micrometre (µm) a unit of measurement equivalent to one-thousandth of a millimetre, or one-millionth of a metre

9780170463027

mid-ocean ridge a broad, high, underwater mountain range in the ocean at a divergent plate boundary

mineral (in food) an inorganic substance present in food that is required by our body to develop and function properly

mineral (in rocks) a naturally occurring inorganic solid with a neatly ordered crystal structure and characteristic composition

mitochondrion an organelle in which energy is released in respiration

mixture a substance made up of two or more different types of particles, physically combined but not chemically bonded

Mohs scale of hardness a scale used to measure the relative hardness and resistance to scratching between minerals

molecular compound a compound formed when two different non-metals chemically bond

molecule the smallest particle of a substance that is capable of separate existence

molecule of a compound a molecule in which two or more atoms of different non-metal elements are chemically bonded together; also known as a molecular compound

molecule of an element a molecule in which two or more atoms of the same non-metal element are chemically bonded together

monatomic/monoatomic an element consisting of just one atom (*mono*– means 'one'; *atomic* means 'atom')

mould a fossil made in the shape of a plant or an animal's remains after those remains have dissolved

multicellular describes a living thing consisting of more than one cell, often many cells

multipotent describes a cell that can differentiate into only a few cell types

nanometre (nm) a unit of measurement equivalent to one-billionth of a metre

nephron the structure in the human kidney where filtration of the blood occurs

neuron a nerve cell

nodules swellings on the roots of plants

non-metal a chemical element that has certain properties, such as being brittle, having a non-shiny appearance and low melting point

non-renewable energy source a source of energy that is finite in nature; i.e. used at a faster rate than it can be produced

nucleus the part of a cell that contains genetic material and is bound by the nuclear membrane

objective lens a lens on a microscope that receives light rays from the specimen and forms an image on the eyepiece

obsidian a volcanic glass

oceanic plate a lithospheric tectonic plate that contains mainly oceanic crust

omnivorous describes an organism that feeds on both plants and animals

opaque describes something that light cannot pass through

ore useful minerals that can be extracted and processed to be sold

organ a collection of different tissues that combine to perform a specific function

organelle a specialised structure in the cytoplasm of a cell that has a specific function

organic relating to, or made from, living material

organoid a small organ grown in the laboratory from stem cells

outer core the liquid part of Earth's metallic core

Pangaea a supercontinent that once existed on Earth; the name means 'all lands'

parent cell the original cell that divides to form two daughter cells

particulates very small pieces of substances suspended in air or in water

period (in chemistry) a horizontal row in the periodic table

periodic table a method of arranging elements by increasing atomic number

peristalsis a progressive wave of contraction and relaxation of the digestive tract

phase change a change in the state of matter; an example of a physical change

phloem plant tissue that transports sucrose from the leaves to the rest of the plant

photosynthesis the process by which a green plant uses sunlight, water and carbon dioxide to produce glucose, which it uses for nutrition, and oxygen, which is released into the air

physical change a change in a substance that does not involve the production of a new substance; can usually be reversed

physical properties the properties of a substance that can be observed or examined without changing its composition

physical weathering a process of weathering that breaks rocks apart or wears them down, but does not change their chemical composition

phytoplankton microscopic, photosynthetic organisms that live in the sea or in fresh water

plasma the watery component of human blood, in which blood cells are suspended

platelets fragments of cells that act in blood clotting

pluripotent describes a cell that has the ability to differentiate into any type of cell in the body

potential energy energy that is stored, ready to be changed to another type of energy

precipitate a solid substance formed in a solution as a result of a chemical reaction

product a new substance produced in a chemical reaction

prokaryote a unicellular organism without a nucleus

property a characteristic or feature of a substance

proteins a complex food group found in foods such as meat, fish, soybeans and cheese

pulmonary artery a blood vessel in humans that takes blood from the heart to the lungs

pulmonary vein a blood vessel in humans that returns blood from the lungs to the heart

pyroclastic describes rock formed from volcanic ash and rock fragments

radioactive element an element that can produce radioactivity

reactant a substance used up in a chemical reaction

red blood cell a blood cell that carries oxygen

regional metamorphism rock metamorphism caused by heat and pressure

relative dating putting fossils, rocks or events into the order in which they occurred

renewable energy source a source of energy that can be produced at a faster rate than it can be used

resolution the finest detail that can be distinguished in an image

ribosome the smallest organelle, which synthesises proteins

ridge push the process that moves crust away from a mid-ocean ridge

rift a valley created by rocks being pulled apart

ring barking a way of killing a tree by removing the bark containing phloem from around the trunk, but leaving the xylem intact

risk factor a condition or behaviour that increases the likelihood of a person developing a disease or health disorder

rock a naturally occurring solid made up of minerals

rock cycle a model used to explain how rocks are formed and how they change

9780170463027

root hair a long extension that provides a large surface area for the root of a plant

root system the water- and nutrient-absorbing part of a plant; usually below ground

rough endoplasmic reticulum (RER) endoplasmic reticulum that produces and transports proteins

science communication talking, raising awareness or even arguing about science-related topics

scientific model a physical, conceptual or mathematical representation of a real phenomenon that is difficult to observe directly

seafloor spreading the theory that new seafloor is created at mid-ocean ridges, spreads outwards and then descends into the mantle at trenches

sediment solid material transported from one place and deposited in another

sedimentary rock rock formed from sediments

septum the dividing wall between the left and right sides of the human heart

shoot system the leaves, stems and flowers of a plant; usually above ground

sieve plate holes at each end of a sieve tube cell that allow the passage of sucrose

sieve tube cell a food-conducting cell in a plant that forms phloem

slab pull the force driving tectonic plate motion caused by a sinking plate

smooth endoplasmic reticulum (SER) endoplasmic reticulum that transports and produces lipids and some carbohydrates

solar power electricity generated directly from sunlight, using solar cells

solute a substance in a solution that is dissolved

solution a mixture produced when a solute dissolves in a solvent

solvent a substance that dissolves another substance to form a solution

sound energy energy transferred as a wave through or by vibrations

specimen a sample to be examined or observed

sphincter a ring of muscle that can close off a tube

stage a flat platform that supports the slide on a light microscope

stain a dye used to colour specimens for microscopic study

starch a complex carbohydrate found in potatoes and other plants; also a form of glucose storage in plants

state of matter one of the forms in which matter can exist: solid, liquid, gas or plasma

stem cell an unspecialised cell able to keep dividing, with the potential to become specialised

stomata (singular: stoma) pores on the surfaces of leaves that allow gas exchange

structural formula a graphic representation that shows the arrangement of atoms in a molecule or compound

subduction the process in which one oceanic plate sinks beneath another plate at a convergent boundary

subliming changing state directly from a solid to a gas without going through a liquid phase

sucrose the form of sugar transported in the phloem of a plant

supercontinent a continent composed of all or most of Earth's continents

surface tension the tension of the surface of a liquid, caused by the attraction of the particles in the surface layer by the bulk of the liquid, which tends to minimise surface area

suspension a mixture of at least one insoluble solid and a liquid or a solution, where the insoluble substance settles to the bottom or rises to the top over time

symptom an indication of a disorder or disease

system a group of organs that work together to perform a specific function

tap root the large tapering main root of some plants

thermal energy energy contained within an object by its vibrating particles, which determines an object's temperature

tidal energy electricity generated from the ebb and flow of the tides

tissue a collection of cells that have similar structures and functions

totipotent describes of a cell that can differentiate into any type of cell or a complete mammal

toxicity a measure of how poisonous a substance is

trace fossil a structure or an impression left by a plant or animal, which shows that life existed

trachea a tube that runs from the back of the throat to the bronchi

transform boundary the border between two tectonic plates that are sliding past each other

transpiration the evaporation of water from the leaves of a plant

transportation the movement of solid particles by agents of erosion over large distances

transuranic an element with an atomic number greater than that of uranium

tsunami a series of large ocean waves created by an undersea earthquake or volcanic eruption

undifferentiated describes a cell that has not yet developed into a specialised cell type

unicellular describes a living thing consisting of a single cell

uplift the process whereby rocks formed underground are raised to the surface

urea nitrogenous waste that is created as amino acids are broken down in the human body

ureter the tube that carries urine from the human kidney to the bladder

urethra the tube that carries urine from the bladder to the outside of the human body

urine liquid containing multiple waste products, especially urea

vacuole a membrane-bound liquid sac found inside a cell

valve a structure in the human heart and veins that prevents backflow of blood

vaporising changing state from a solid or a liquid to a gas; e.g. evaporating, boiling or subliming

vascular bundle a combined strand of xylem and phloem tissue in plants

vein (in humans) a blood vessel in humans that carries blood to the heart

vein (in plants) a vascular bundle of xylem and phloem tissue in a leaf

vena cava a large vein in humans that brings blood to the heart from all parts of the body

vent the central opening of a volcano

ventricles the two lower chambers in the human heart, which pump blood to either the lungs or the rest of the body

villi small finger-like projections on the cells of the small intestine that increase surface area

viscosity a measure of a fluid's resistance to flow

vitamin a substance that is essential in very small amounts for normal growth and activity

volcano an opening in Earth's crust, through which molten rock reaches the surface

waste unwanted or unusable material, substances or by-products

waste energy energy that is a result of energy transformation, but that is not usable within the energy transformation system

weathering the breakdown of minerals and rocks at Earth's surface by physical or chemical processes

wet mount a glass slide that holds a specimen in a liquid such as water for viewing under a microscope

white blood cell a blood cell that is part of the human immune system

wilt to become limp and to droop through loss of water

wind power electricity generated by harnessing the kinetic energy of the wind to drive a turbine

word equation a word summary that shows the reactants and products of a chemical reaction

work energy that is being transferred or transformed

xylem plant tissue that transports water and minerals from the roots to the rest of the plant

Index

A

absolute dating 168

acid mine water 173

adult stem cell 22

alveoli 43

amoeba 18

animals

    body organisation 34–5

    cells 14–15, 34, 35

    digestive system 40–1

    internal systems of 52–3

    requirements of 32–3, 63

anus 39

aorta 45

appendix 39

artefact 166

    manufacturing 171–2

arteries 46

asthenosphere 184

atherosclerosis 50

atmosphere 182

atomic weight 94

atoms 86–93

    changing concept of 90–3

atria 44, 45

axon 19

B

bark, harvesting 77–8

batholith 151

bed 158

biconcave 19

biodegradable materials 136

biodiversity 79

biofuels 228

biological weathering 156

biosphere 183

blood 44–7

    circulation 44–5

    vessels 46

body fossil 166

body organisation, animals 34–5

body systems, disorders of 50–1

boiling 120

bolus 37

bowel motion 39

breathing 42

    expiration 42

    inspiration 42

bronchi 42

bronchioles 43

bushfires 76

C

caecum 40

calorie (cal) 214

canopy 74

capillaries 46

carbohydrates 36

carbon dioxide 62

cardiovascular system 44–7

carnivorous 40

cartilage 42

cast 167, 168

cell theory 5

cells 4–23

    animal 14–15, 34, 35

    daughter 5

    defined 4

    discovery of 4–5

    guard 21, 69, 72–3

    membrane 10–11

    parent 5

    plant 16–17

    size of 6–7

    specialised 18–21

    stem 22–3

    wall 16, 17

cellular respiration 14

cellulose 16

cementation 158

chambers, heart 44, 45

charcoal 134

chemical bond 98

chemical change 122–7

    defined 122

    evidence of 125–7

    examples of 123

    indicators of 125–7

    particle models for 124

chemical digestion 36

chemical equation 123

chemical formula 102, 103

chemical potential energy 128, 219

chemical properties 131

chemical reaction 122, 123

    energy in 128

chemical symbol 94–5

chemical weathering 155

chlorophyll 16, 64

chloroplasts 16, 17, 64

cholesterol 50

chyme 37

clastic sedimentary rock 159

cleavage 148, 149

coarse focus knob 9

colloid 106

colour 126, 148

compaction 158

companion cell 71

compound
  chemical formula 102
  defined 99
  ionic 99
  molecular 98, 99
concentration 11, 105
condensing 120
contact metamorphism 160
continental drift 192–3
continental plate 185
convection 186
  from cold surface 187
  continental drift and 193
convergent boundary 188
copper 86
coronary artery disease 50–1
corrode 131
crust of Earth 183
crystal 147
cuticle 72
cycads 78
cytoplasm 4

D

Dalton, John 90, 91, 92–3
daughter cells 5
Democritus 90, 91, 92
density 94, 148
deposition 156
diagnosis 50
diaphragm 42
diatomic elements 100
differentiation 19
diffusion 11
digestive system 36–41
  animals 40–1
  anus 39
  large intestine 39
  mouth 36–7

oesophagus 37
  rectum 39
  small intestine 38
  stomach 37–8
digestion
  chemical 36
  mechanical 36
divergent boundary 188
doping 108
drug testing in sport 108

E

Earth
  atmosphere 182
  biosphere 183
  continental drift 192–3
  crust 183
  geosphere 182, 183
  hydrosphere 182–3
  inner core 183
  outer core 183
  mantle 183
  plate boundaries 188–91
  plate tectonics 184–7
  structure of 182–3
earthquakes 190, 198, 199–200
  First Nations cultural
    narratives of 203
eclogite 160
egest 39
elastic potential energy 220
electrical energy 216
  energy transformation into
    227–9
  transferring 224
electricity
  generating in remote
    communities 233
electrons 24
elements 88–9

chemical formula 102
  developing and refining
    concept of 92–3
  diatomic 100
  molecule of 98
  monatomic/monoatomic 100
Elements of Chemistry (Lavoisier) 92
embryonic stem cell 22
endoplasmic reticulum 14
endoscope 50
energy
  changes in rock cycle 164–5
  defined 214
  from food 32
  heat as form of 128
  in chemical reaction 128
  kinetic 214–17, 223–4
  potential 218–21
  transformation 226–9
energy transfer 222–5
  defined 222
  examples of 223–4
  law of conservation of
    energy 222
  modelling with diagrams 225
energy transformation 226–9
  by friction 229
  defined 226
  examples of 226
  into electrical energy 227–9
  modelling with diagrams 227
enzymes 14
ephemeral 76
epidermis 72
erosion 156–7
  deposition 156
  transportation 156
estuary 74
eukaryotes 12–13
evaporating 120
excretion 48

expiration 42

extrusive igneous rocks 151

eyepiece lens 8

F

fat cells 19

fats 36

fermentation 133

fibre 39

fibrous roots 67

field of view 8

filtration 48–9

fine focus knob 9

fire-making techniques

    fire saw method 231–2

    hand drill method 230–1

    traditional tools and methods of 230–2

fire saw method of fire making 231–2

First Nations Australians

    body systems, knowledge of 52–3

    plants, knowledge and use of 77–8

    records of geological events in Australia 202–3

    shaping of rocks into tools 170–2

    traditional fire-making techniques 230–2

    use of chemical and physical changes 133–5

fissures 199

flammability 131

foliation 162

food 32–3

    labels 220

fossils 166–9

    age of 168–9

    body 166

    formation 167–8

    modelling 168

    trace 166

freezing 119

friction

    defined 229

    energy transformation by 229

G

gangue minerals 149

gases 126

    control in plants 72–3

gas exchange 43

    in leaf 72–3

geosphere 182, 183

geothermal energy 228

Golgi body 14

graphite 148

gravitational potential energy 218

greenhouse gases 227

group, in periodic table 95–6

guard cells 21, 69, 72–3

H

haemoglobin 46–7

hand drill method of fire making 230–1

hard parts of organisms 166

heart 44–5

heartburn 37

heat

    and thermal energy 215–16

    as form of energy 128

    interpreting data for change in 128–9

Helicobacter pylori 51

herbivorous 40

Hooke, Robert 4

hormones 46, 47

hornfels 161

hotspot, volcanic 191

humans

    cardiovascular system of 44–7

    digestive system of 36–9

    organ systems of 35

    respiratory system of 42–3

    urinary system of 48–9

hydrochloric acid 37

hydroelectricity 228

hydrosphere 182

hydrothermal metamorphism 163

I

igneous rocks 150–3

    classified by composition and texture 153

    extrusive 151

    formation of 150–2

    identifying 152–3

    intrusive 151

immune system 47

induced pluripotent stem cell 23

ingested 36

ink 106

inner core of Earth 183

inorganic 146

insoluble 125, 126

inspiration 42

intrusive igneous rocks 151

ionic compound 99

iris diaphragm 9

J

Janssen, Hans 4

Janssen, Zacharias 4

joule (J) 214

K

kidneys 48–9
  function 48
  stones 50
kinetic energy 214–17, 223–4
  defined 214
  transferring 223–4
  types of 215–17

L

large intestine 39
lava 150
Lavoisier, Antoine 92
law of conservation of energy 222
leaf
  adaptations to hot, dry
    environments 75
  gas exchange in 72–3
  movement of water
    through 73
  size and type 75
  tissues of 72
light energy 217
light microscope 8–9
lignin 68
lignotuber 76
lithification 167
lithosphere 184
lustre 148, 149
lysosomes 14

M

magma 150
magnetic reversal 196, 197
magnification 8, 9
mantle of Earth 183
Mars 145
mechanical digestion 36
mechanical energy 221

melting 119
membrane 4
Mendeleev, Dmitri 94
meniscus 137
metal 88, 89
metalloid 88
metamorphic rocks 160–3
  defined 160
  examples of 163
  features of 162
  metamorphism, types of 160–1
micrograph 24
micrometres 6
microscopes 4, 6–9, 24
  light 8–9
  magnification 8, 9
microscopy techniques 9
mid-ocean ridge 186
minerals
  defined 32, 146
  identification of 148
  rocks and 146–9, 170–1
  structure 147–8
mining 170–1
mitochondria 14
mixture
  defined 104
  types of 104–6
Mohs scale of hardness 148, 149
molecule 98
molecule of a compound (molecular
  compound) 98, 99
molecule of an element 98
monatomic/monoatomic
  elements 100
mould 167
mouth 36–7
multicellular animals 19–21
  specialisations in 19
multicellular organisms
  10, 34

multipotent stem cells 22
muscle cells 19

N

nanometres 6
nephrons 48
neurons 19
*A New System of Chemical
  Philosophy* (Dalton) 93
nodules 63
non-metal 88
non-renewable energy source 227
nucleus 11

O

objective lens 8
obsidian 151
ocean floor, discovering 204
oceanic plate 185
oesophagus 37
omnivorous 40
opaque 106
ore 149
organelles 4
organic sedimentary rock 159
organoids 54
organs 34
organ system 35
  human 35
outer core of Earth 183
oxygen 62

P

Pangaea 192
parent cells 5
particulates 217
Pelagibacter ubique 4
peptic ulcers 51

period, in periodic table 95–6

periodic table 94–6

    modern 95–6

    original 94–5

peristalsis 37

phase change 119–20

phloem 70

    structure of 70–1

photosynthesis 16, 64–5

    respiration vs 65–6

    site of 64

physical change 118–21

    by mixing 120–1

    changes of shape 118

    defined 118

    phase change 119–20

    state of matter 118–19

physical properties 131

physical weathering 155

phytoplankton 65

plant cells 16–17

    specialised 20–1

plants

    adaptations to fire 76

    adaptations to hot, dry environments 75

    control of gases 72–3

    inputs and outputs of 62

    minerals required by 63

    photosynthesis 64–5

    rainforest 74–5

    requirements of 62–3

    specialised structures in 74–6

    sugar transport 70–1

    sustainably harvested materials 77–8

    systems in 66–7

    water transport 68–9

plasma 46, 47

plaster 134

plate boundaries 188–91, 198

platelets 47

plate tectonics 184–91, 204

    evidence for 190–1

    history of 204

pluripotent stem cell 22

potential energy 218–21

    defined 218

    types of 218–21

precipitate 125

product 122

prokaryotes 11–12

properties 88

proteins 36

pulmonary artery 45

pulmonary vein 45

pure substance 87

pyroclastic 151

pyroligneous acid 134

pyrolysis 133–4

Q

quarrying 170–1

quartzite 161

R

radioactive element 168

rainforest plants 74–5

reactant 122

rectum 39

red blood cells 19, 46, 47

regional metamorphism 163

relative dating 168

remote communities, generating electricity in 233

renewable energy source 228

resins 133

resolution 24

respiration vs photosynthesis 65–6

respiratory system 42–3

ribosomes 14

ridge push 186

rift 195

ring barking 71

risk factors 51

rock cycle 164–5

rocks

    defined 146

    igneous 150–3

    metamorphic 160–3

    and minerals 146–9, 170–1

    sedimentary 158–9

    testing for hardness and strength 172

    see also rock cycle

root hair 20, 68–9

root system 67, 75

roots

    adaptations to hot, dry environments 75

    fibrous 67

    tap 67

rough endoplasmic reticulum (RER) 14

Rube Goldberg machine 213, 237

S

saliva 36–7

salt 85

science communication 136

scientific model 92

seafloor spreading 194–7

    evidence for 195–7

sediment 156

sedimentary rocks

    defined 158

    types of 159

septum 45

shoot system 66

sieve plate 70

sieve tube cell 70

slab pull 186–7

small intestine 38

smooth endoplasmic reticulum (SER) 14

solar power 228

solute 105

solutions 104–6

solvent 105

sound energy 215

specialised cells 18–21

specimen 8

sphincter 37

stage 8

stain 9

starch 37

states of matter 118–19

stem cells 22–3

    adult 22

    embryonic 22

    multipotent 22

    pluripotent 22

    totipotent 22

stomach 37–8

stomata 21, 72–3

    adaptations to hot, dry environments 75

stones, kidney 50

structural formula 103

subduction 188

subliming 120

substances

    chemical properties 131

    physical properties 131

    properties affect use of 132

sucrose 70

sugar transport in plants 70–1

sunlight 62

supercontinent 192

surface tension 137

suspension 106

symptom 50

T

tap roots 67

temperature 127

thermal energy 215–16

3D models of atoms 102–3

tidal energy 228

tissues 34

    of leaf 72

totipotent stem cells 22

toxicity 131

trace fossil 166

trachea 42

transform boundary 188

transpiration 69

transportation 156

transuranic 88

tropical rainforests 117

tsunamis 201

    First Nations cultural narratives of 203

U

ulcers, peptic 51

umbilical cord blood 3

undifferentiated cells 22

unicellular organisms 10, 18, 34

    specialisations in 18

uplift 165

urea 48

ureter 49

urethra 49

urinary system 48–9

urine, formation of 48–9

V

vacuoles 14

valve 46, 47

van Leeuwenhoek, Antonie 4, 5

vaporising 120

vascular bundles 68

veins 46, 68

vena cava 45

vent 199

ventricles 44, 45

villi 38

viruses 12

viscosity 198

vitamin 32

volcanoes 150, 198–9

    First Nations cultural narratives of eruptions 202–3

    hotspot 191

W

waste 32

    removal 48

waste energy 229

water 32

    conducting tissues 68

    requirements of plants 62

    transport in plants 68–9

    values for intake and output in humans 48

weathering 154–6

    biological 156

    chemical 155

    defined 154

    physical 155

wet mount 9

white blood cells 20, 47

wind power 228

wood ash 134

word equation 123

work 214

X

xylem 68

Z

zinc 86, 87

Additional credits

Chapter 1

▶ **page 6 (Table 1.2.1):**   first row: Shutterstock.com/Pineapple studio

▶ **page 7 (Table 1.2.1):**   top to bottom: Shutterstock.com/Lebendkulturen.de, Shutterstock.com/Nemes Laszlo, Getty Images/Stone/Ed Reschke, Shutterstock.com/Digital Photo, Shutterstock.com/Phonlamai Photo, Getty Images Plus/iStock/spawns

Chapter 2

▶ **page 47 (Table 2.5.2):**   top to bottom: Getty Images Plus/iStock/Kalinovskiy, Alamy Stock Photo/BSIP SA, Alamy Stock Photo/Science Photo Library, Alamy Stock Photo/Science Photo Library

Chapter 4

▶ **pages 90–91 (timeline):**   top left to right: Wellcome Collection. Public Domain Mark, Alamy Stock Photo/SPCOLLECTION, Alamy Stock Photo/Pictorial Press Ltd, Alamy Stock Photo/Science History Images; bottom left to right: Getty Images/De Agostini; Getty Images/Archive Photos; Alamy Stock Photo/Painters; Getty Images/Hulton Archive/Imagno

▶ **page 96 (Table 4.3.1):**   Column 2, top to bottom: Shutterstock.com/Wojmac, Shutterstock.com/Tatiana Ivleva, Shutterstock.com/macrowildlife; column 3, top to bottom: Shutterstock.com/Rvkamalov gmail.com, Shutterstock.com/Sebastian Janicki, Shutterstock.com/Fablok; column 4, top to bottom: Alamy Stock Photo/Science Photo Library, Shutterstock.com/Bjoern Wylezich, Getty Images Plus/E+/AndreasKermann

Chapter 5

▶ **page 119 (Table 5.1.1):**   top to bottom: Shutterstock.com/photoschmidt, Getty Images/iStock/DonNichols

▶ **page 120 (Table 5.1.1):**   top to bottom: Shutterstock.com/fotopai, Getty Images/EyeEm/Duaa Awchi, Getty Images/iStock/RBOZUK

▶ **page 123 (Table 5.2.1):**   top row left to right: Shutterstock.com/freedarst, Getty Images Plus/iStock/Birdlkportfolio; middle row left to right: Shutterstock.com/Arina P Habich, Shutterstock.com/Denise E; bottom row left to right: Shutterstock.com/Tatiana Popova, Shutterstock.com/Jacob_09

▶ **page 127 (Question 4):**   a: Shutterstock.com/Natasa Re; b: Science Photo Library/SCIENCE SOURCE/CHARLES D. WINTERS; c: Shutterstock.com /Christos Siatos; d: Science Photo Library/TURTLE ROCK SCIENTIFIC/SCIENCE SOURCE; e: Shutterstock.com/PHOTO FUN

Chapter 8

▶ **page 226 (Table 8.4.1):**   top to bottom: Shutterstock.com/k_samurkas; Shutterstock.com/cosma; Gettyimages/iStock/the-lightwriter